Nordic Botanical Research in the Andes and Western Amazonia

edited by
S. Lægaard and F. Borchsenius

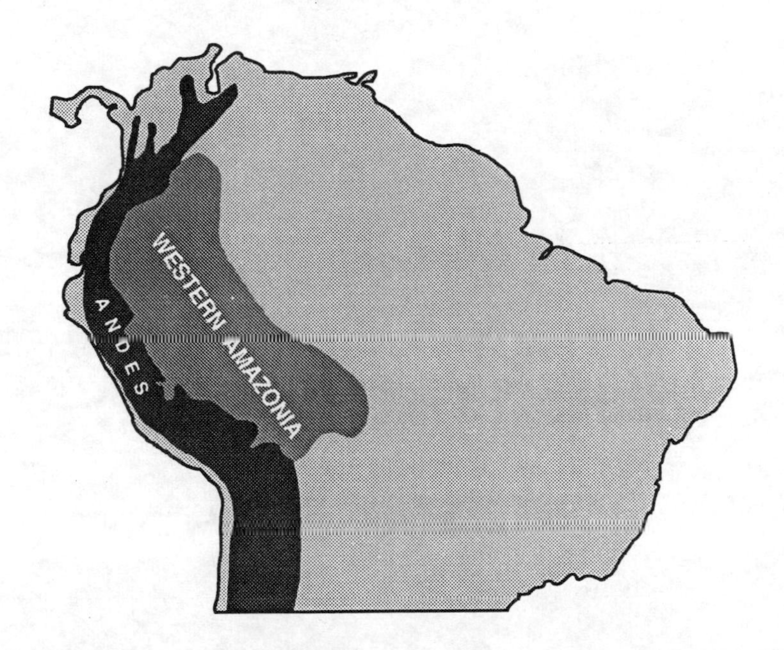

AAU REPORTS 25

Botanical Institute Aarhus University 1990
this issue in collaboration with
Pontificia Universidad Católica del Ecuador, Quito

Contents

Contents

Introduction

A symposium entitled "Nordic Botanical Research in Tropical South America" was held 28–30 August 1989 at the congress center Sandbjerg, of Aarhus University.

Similar meetings have been held on two previous occasions: in 1984 (Reports Bot. Inst. Univ. Aarhus no. 9, 1984) and in 1986 (Reports Bot. Inst. Univ. Aarhus no. 15, 1986). At both of these meetings staff members and advanced students from the universities of Göteborg and Aarhus participated.

This 3rd meeting was organized according to a newly established agreement of collaboration between Aarhus University, University of Bergen, University of Göteborg, University of Turku, and Åbo Academy. Participants came from most of these and from a few other Nordic universities.

We wish to express our gratitude to Nordiska Forskarkursar for financial support to the symposium. We also wish to express our gratitude to the many collaborators in South America, whose collaboration and active support have been essential for the results of our work. We are especially grateful for the help from the following Ecuadorian institutions: Dept. de Ciencias Biológicas, P. Universidad Católica del Ecuador; Dept. de Parques Nacionales y Vida Silvestre, Ministerio de Agricultura y Ganadería; CONUEP; Museo Ecuatoriano de Ciencias Naturales, Quito; Muséo Antropologico del Banco Central, Guayaquil; PREDESUR, Loja; Dept. de Botánica, Fisiología y Sistemática, Universidad Nacional de Loja.

March 1990

Simon Lægaard
Finn Borchsenius

List of contributors to *Flora of Ecuador*

Gunnar Harling

Botaniska Institutionen, University of Göteborg, Carl Skottsbergs Gata 22, S-413 19 Göteborg, Sweden

FAMILY	AUTHOR	INSTITUTION	PUBL.
Lycopodiaceae	B. Øllgaard	AAU	1988
Selaginellaceae	C. A. Jermy & B. Øllgaard	BM, AAU	
Isoetaceae	J. Hickey	KY	
Psilotaceae	B. Øllgaard	AAU	
Equisetaceae	B. Øllgaard	AAU	
Ophioglossaceae	B. Øllgaard	AAU	
Osmundaceae	B. Øllgaard	AAU	
Schizaeaceae	B. Øllgaard	AAU	
Hymenophyllaceae	P. Windisch & B. Øllgaard	HB, AAU	
Loxomataceae	B. Øllgaard	AAU	
Dicksoniaceae	R. Tryon	GH	1986
Lophosoriaceae	R. Tryon	GH	1986
Metaxyaceae	R. Tryon	GH	1986
Cyatheaceae	R. Tryon	GH	1986
Polypodiaceae			
Pteridoideae	A. Luz Arbeláez	HUA	
Thelypteridoideae	A. R. Smith	UC	1983
Asplenioideae	R. G. Stolze	F	1986
Blechnoideae	B. Øllgaard	AAU	
Marsileaceae	B. Øllgaard	AAU	
Salviniaceae	B. Øllgaard	AAU	

FAMILY	AUTHOR	INSTITUTION	PUBL.
Cycadaceae	D. Stevenson	NY	
Pinaceae	D. Stevenson	NY	
Cupressaceae	D. Stevenson	NY	
Podocarpaceae	D. Stevenson	NY	
Myricaceae	S. Lægaard & U. Molau	AAU, GB	
Juglandaceae	S. Lægaard & U. Molau	AAU, GB	
Salicaceae	S. Lægaard & U. Molau	AAU, GB	
Betulaceae	S. Lægaard & U. Molau	AAU, GB	
Cecropiaceae	C. C. Berg	BG	
Moraceae	C. C. Berg	BG	
Eremolepidaceae	J. Kuijt	LEA	1986
Viscaceae	J. Kuijt	LEA	1986
Loranthaceae	J. Kuijt	LEA	1986
Balanophoraceae	B. Hansen	C	1983
Polygonaceae	J. Brandbyge	AAU	1989
Phytolaccaceae	U. Eliasson	GB	
Achatocarpaceae	U. Eliasson	GB	
Nyctaginaceae	J.-E. Bohlin	GB	
Molluginaceae	U. Eliasson	GB	
Aizoaceae	U. Eliasson	GB	
Portulacaceae	U. Eliasson	GB	
Basellaceae	U. Eliasson & R. Eriksson	GB	
Caryophyllaceae	U. Eliasson	GB	
Chenopodiaceae	U. Eliasson	GB	
Batidaceae	U. Eliasson	GB	
Amaranthaceae	U. Eliasson	GB	1987
Cactaceae	J. Madsen	AAU	1989
Monimiaceae	S. Renner	AAU	
Lauraceae	H. van der Werff	MO	
Berberidaceae	C. Ulloa	QCA	
Nymphaeaceae	L. B. Holm-Nielsen	AAU	
Ceratophyllaceae	L. B. Holm-Nielsen	AAU	
Piperaceae	R. Callejas	HUA	
Chloranthaceae	C. Todzia	TEX	
Actinidiaceae	D. D. Soejarto	F	1982
Caryocaraceae	G. T. Prance & S. Mori	K, NY	

FAMILY	AUTHOR	INSTITUTION	PUBL.
Marcgraviaceae	H. G. Bedell	GH	
Clusiaceae	B. Maguire	NY	
Hypericaceae	N. K .B. Robson	BM	
Papaveraceae	M. Lidén	GB	
Capparidaceae	H. H. Iltis	WIS	
Cruciferae	B. Jonsell	SBT	
Cunoniaceae	G. Harling	GB	
Rosaceae	K. Romoleroux	QCA	
Chrysobalanaceae	G. T. Prance	K	1979
Connaraceae	E. Forero	MO	
Mimosaceae	I. Nielsen, coordinator	AAU	
Caesalpiniaceae	D. Neill	MO	
Papilionaceae	I. Nielsen, coordinator	AAU	
Tropaeolaceae	B. Sparre†	S	1973
Erythroxylaceae	T. Plowman†	F	1989
Euphorbiaceae	M. Huft, coordinator	F	
Simaroubaceae	W. Thomas	NY	
Burseraceae	D. Daly	NY	
Malpighiaceae	W. Anderson	MICH	
Polygalaceae	B. Eriksen	GB	
Coriariaceae	L. E. Skog	US	1987
Anacardiaceae	A. Barfod	AAU	1987
Rhamnaceae	M. C. Johnston	TEX	
Malvaceae	P. Fryxell	TEX	
Sterculiaceae	L. Dorr	NY	
Dichapetalaceae	G. T. Prance	K	1980
Violaceae	U. Molau & H.A. Hekking	GB, U	
Passifloraceae	L. B. Holm-Nielsen, J. Lawesson, & P. M. Jørgensen	AAU	1988
Bixaceae	U. Molau	GB	1983
Cochlospermaceae	U. Molau	GB	1983
Elatinaceae	U. Molau	GB	1983
Caricaceae	V. M. Badillo	MY	1983
Loasaceae	M. Poston	MO	
Begoniaceae	L. B. Smith & D. Wasshausen	US	1986

FAMILY	AUTHOR	INSTITUTION	PUBL.
Cucurbitaceae	R. P. Wunderlin	USF	
Lythraceae	A. Lourteig	P	1989
Lecythidaceae	G. T. Prance & S. Mori	K, NY	
Melastomataceae	J. J. Wurdack	US	1980
Rhizophoraceae	L. Andersson	GB	
Combretaceae	C. Stace	MANCH	
Onagraceae	P. A. Munz†	UC	1974
Umbelliferae	M. E. Mathias & L. Constance	LA, UC	1976
Clethraceae	C. Gustafsson	GB	
Ericaceae	J. Luteyn	NY	
Theophrastaceae	B. Ståhl	GB	1990
Primulaceae	B. Ståhl	GB	1990
Plumbaginaceae	J. Luteyn	NY	1990
Styracaceae	R. Monteiro	HRCB	
Symplocaceae	B. Ståhl	GB	
Oleaceae	B. Ståhl	GB	
Gentianaceae	J. S. Pringle	HAM	
Asclepiadaceae	G. Morillo	VEN	
Rubiaceae	L. Andersson, coordinator	GB	
Convolvulaceae	D. F. Austin	FAU	1982
Boraginaceae	H. Riedl	W	
Labiatae	R. M. Harley	K	
Solanaceae	W. G. D'Arcy, coordinator	MO	
Buddlejaceae	E. M. Norman	FTU	1982
Scrophulariaceae	N. H. Holmgren & U. Molau	NY, GB	1984
Bignoniaceae	A. H. Gentry	MO	1977
Acanthaceae	D. Wasshausen	US	
Gesneriaceae	L. E. Skog & L. P. Kvist	US, AAU	
Columelliaceae	K. Fagerström	S	1975
Lentibulariaceae	P. Taylor	K	1975
Plantaginaceae	K. Rahn	C	1975
Valerianaceae	B. Eriksen	GB	1989
Campanulaceae	S. Jeppesen	AAU	1981
Lobeliaceae	S. Jeppesen	AAU	1981
Sphenocleaceae	S. Jeppesen	AAU	1981
Goodeniaceae	S. Jeppesen	AAU	1981

FAMILY	AUTHOR	INSTITUTION	PUBL.
Compositae			
Vernonieae	S. Keeley	CONN	
Liabeae	H. E. Robinson	US	1978
Eupatorieae	H. E. Robinson & R. M. King	US	
Heliantheae	H. E. Robinson	US	
Anthemideae	G. Harling	GB	
Senecioneae	B. Nordenstam & R. Lundin	S	
Mutisieae	G. Harling	GB	
Alismataceae	L. B. Holm-Nielsen & R.R. Haynes	AAU, UNA	1986
Limnocharitaceae	R. R. Haynes & L. B. Holm-Nielsen	UNA, AAU	1986
Hydrocharitaceae	R. R. Haynes	UNA	1986
Juncaginaceae	R. R. Haynes & L .B. Holm-Nielsen	UNA, AAU	1986
Potamogetonaceae	R. R. Haynes & L. B.Holm-Nielsen	UNA, AAU	1986
Zannichelliaceae	R. R. Haynes & L. B. Holm-Nielsen	UNA, AAU	1986
Najadaceae	R. R. Haynes & L. B. Holm-Nielsen	UNA, AAU	1989
Liliaceae	P. Ravenna	SGO	
Haemodoraceae	P. J . M. Maas	U	subm.
Alstroemeriaceae	M. Neuendorf & U. Molau	GB	
Amaryllidaceae	A. Meerow	FLAS	subm.
Pontederiaceae	C. Horn	UNA	1987
Iridaceae	P. Ravenna	SGO	
Burmanniaceae	P. J. M. Maas	U	
Juncaceae	H. Balslev	AAU	1979
Bromeliaceae	H. Luther	SEL	
Commelinaceae	D. R. Hunt	K	
Graminae	S. Lægaard, coordinator	AAU	
Bambusoideae	E. Judziewecz & L. G. Clark	US, TEX	
Centothecoideae	S. Lægaard	AAU	
Arundinoideae	S. Lægaard	AAU	
Chloridoideae	S. Lægaard	AAU	

FAMILY	AUTHOR	INSTITUTION	PUBL.
Pooideae	S. Lægaard	AAU	
Panicoideae	S. Renvoize	K	
Palmae	H. Balslev	AAU	
Cyclanthaceae	G. Harling	GB	1973
Araceae	T. B. Croat	MO	
Typhaceae	L. B. Holm-Nielsen	AAU	
Cyperaceae	T. Koyama	NY	
Musaceae	L. Andersson	GB	1985
Zingiberaceae	P. J. M. Maas	U	1976
Cannaceae	P. J. M. & H. Maas	U	1988
Marantaceae	H. Kennedy, L. Andersson, & M. Hagberg	UBC, GB	1988
Orchidaceae			
Cypripedioideae-Neottioideae	L. A. Garay	AMES	1978
Remain. subfam.	C. Dodson, coordinator	MO	

The history of botanical exploration in Amazonian Ecuador

Susanne S. Renner

Botanical Institute, Aarhus University, Nordlandsvej 68, DK 8240 Risskov, Denmark

The floristic exploration of Ecuador began with the great exploring expeditions of Charles Marie de la Condamine and Joseph de Jussieu (1739–48; 1743) and continued in the 19th century with an uninterrupted stream of illustrious visitors passing the country on their way from Bogotá to Lima.

However, the first botanical collections coming undoubtedly from the eastern lowlands below 600 m altitude were made by Jameson who in 1857 travelled from Archidona to Tena, Puerto Napo, Ahuano, Sta. Rosa and back. The next explorers, in chronological sequence, were Spruce, Isern y Battló, Wallis, Poortman, Rimbach, Crespi, Benoist, Heinrichs, Mexía, and the Schultze-Rhonhofs. Only some 20 people appear to have botanized in the lowlands during the first 100 years of botanical history until 1958. The number of collectors was about the same during the next 10 years and rose to 140 during the last 18 years (up to May 1988). Most collectors are still alive, and it has thus been possible to obtain "first hand" information from them via letters and interviews on itineraries, dates, and numbers of plants collected. Several important collections were made by ethnobotanists such as Schultes, Pinkley, Vickers, Davis and Yost, Alarcón, Baker and Lowell, and Irvine; unfortunately, but understandably, their collections usually consist of few duplicate sets.

What does a history of collecting in Ecuadorean Amazonia tell us besides some people's age and initials?

First, besides giving a quick overview of who has been where, how much they collected, and where the collections are, it shows that ca. 61,000 specimens, *i.e.*, about 52 specimens per 100 km^2 have so far been gathered in the eastern lowlands (118,000 km^2). Thus the number given in a recent book on *Tropical Inventories* (Campbell & Hammond 1989) by Campbell of 21 specimens per 100 km^2 for Ecuador as a whole based on figures from the *Index Herbariorum* (1981) is now outdated. An extraordinary increase in collecting has occurred during the last years, for example, due to Cerón (ca. 5,000 nrs.), Palacios *et al.* (3,800 nrs.), Brandbyge *et al.* (4,265 nrs.), Neill *et al.* (3550 nrs.), Balslev *et al.* (1,890 nrs.), M. Baker (1,800 nrs.), Holm-Nielsen *et al.* (1,757 nrs.), Øllgaard *et al.* (1,696 nrs.), and others. Collecting is, however, still very uneven.

References

Campbell, D. G. and Hammond, H. D. (eds.) 1989. Floristic inventories of tropical countries. – New York Botanical Garden, New York.

Neotropical Rubiaceae: a project in *statu nascendi*

Lennart Andersson

Botaniska Institutionen, University of Göteborg, Carl Skottsbergs Gata 22, S-413 19 Göteborg, Sweden

The Rubiaceae comprise some 225 genera and 5,000 species in the Neotropics. Most species are shrubs and small trees; tall trees, subshrubs, vines, epiphytes, and herbs are rather poorly represented. The family forms a conspicuous element in all Neotropical biota, except in the high Andean ones, and in the arid lowlands. The largest number of species, by far, is to be found in lowland rainforest and low to mid elevation montane forests.

In view of its size and ecological importance, the family is surprisingly poorly known taxonomically. Only half a dozen genera or so have been monographed in modern times. Most of our knowledge at the species level is scattered in the floristic literature. Standley, by far the most productive researcher on this family, treated Central America and the Caribbean in the North American Flora (1918–34) and all the Andean states from Venezuela to Bolivia in a series of papers in the Field Museum Publications (1930–31). Steyermark treated all the Rubiaceae of the Guiana Highlands in the Memoirs of the New York Botanical Garden (1964–1972) and also treated the family for the Flora de Venezuela. The family has also been treated in Flora Brasiliensis (Müller 1881, Schumann 1888), Flora of Suriname (Bremekamp 1934), and Flora of Panama (Dwyer 1980).

The gross taxonomy is poorly understood. Many genera are vaguely circumscribed and hypotheses about their interrelationships are based on few characters and dubious homologies. In spite of three ambitious attempts to revise the suprageneric classification (Verdcourt 1958, Bremekamp 1966, and Robbrecht 1988), we are nowhere close to a consensus about the precise circumscription of subfamilies and tribes. No attempt has been made to try numerical approaches, neither numerical cladistics, nor numerical phenetics. In fact, even Robbrecht's approach is essentially a key character one.

The "Gothenburg Rubiaceae Project" takes its starting point in a need to find an "umbrella" project in which to engage an entire research group. Groups are generally more successful in fund-raising than individual researchers and offer a more stimulating scientific environment than institutes where each one is absorbed by his own isolated project. Both points are important for the success in graduate training programs.

The following points were considered when looking for a suitable taxon: 1) it must be large enough to have room for half a dozen people or so, for a reasonable length of time; 2) there should not be too many people elsewhere already working on it; 3) there must be a number of readily circumscribed subgroups of a size suitable for Ph. D. projects; 4) there should be need for and promise of real progress; 5) the kind of work needed should make good use of preexisting competence within the institute; 6) the data generated in subprojects should be useful in a broader theoretical framework common to the whole group. The Rubiaceae was the family that seemed to me best to meet these demands in the neotropical flora. Luckily, it is also one that appeals to me from the important, purely irrational point of view; I like it.

In order to get a general idea about the family a preliminary reconnaissance was made in some larger herbaria: C, F, G, MO, NY, P, S, and US. All species were recorded together with notes about their occurrence in the 19 major phytogeographical regions of the Neotropics. The result was a rough geographic checklist of all neotropical Rubiaceae, which will eventually be computerized. As a first test, the genera of the tribe Cinchoneae were worked up. The test confirms the suspicion that the checklist material is indeed quite helpful when trying to identify genera suitable for further study.

The plans for the immediate future are first to go on with the working up of the checklist material, a task which I hope can be largely done by a secretary, once the formate and routines are set. Secondly, a reevaluation of the suprageneric classification of the subfamily Cinchonioideae will be made, using numerical methods. This second line of inquiry will be made as a B. Sc. thesis work. Funding to go on with the project has been granted by the Swedish Natural Science Research Council for the next two years.

The unifying theoretical framework will be historical biogeography. Recent reevaluations of the fossil record challenge the view that tropical biota have existed unchanged since early Cretaceous. Instead, it appears that mesothermic forests are "older" than all tropical biota, which seem to have arisen gradually during the Paleocene. If the phylogeny of a large and ecologically diversified group, such as the Rubiaceae, could be reconstructed with reasonable certainty, it would cast light on the history of the biomes in which the plants occur. To try to do so, is the ultimate, superior goal of the "Gothenburg Rubiaceae Project".

References

Bremekamp, C. E. B. 1934. Rubiaceae. – In: A. Pulle (ed.), Flora of Suriname 4(1): 113–298. Reprint: Brill, Leiden, 1966.

– 1966. Remarks on the position, the delimitation and the subdivision of the Rubiaceae. – Acta Bot. Neerl. 15: 1–33.

Dwyer, J.D. 1980. Rubiaceae. – In: R.E. Woodson & R.W. Schery (eds.), Flora of Panama IX. Ann. Missouri Bot. Gard. 67: 1–522.

Müller, J. (Argoviensis) 1881. Rubiaceae. – In: C.F.P. von Martius, A.W. Eichler & I. Urban (eds.), Flora Brasiliensis 6(5). – Leipzig.

Robbrecht, E. 1988. Tropical Woody Rubiaceae.– Opera Bot. Belgica 1.

Schumann, K. 1889. Rubiaceae. – In: C.F.P. von Martius, A.W. Eichler & I. Urban (eds.), Flora Brasiliensis 6(6). Leipzig.

Standley, P.C. 1918–1934. Rubiaceae. – North American Flora 32(1–4). New York Botanical Garden, New York.

– 1930. The Rubiaceae of Colombia. – Publ. Field. Mus. Nat. Hist., Bot. Ser. 7: 1–176.

– 1931a. The Rubiaceae of Ecuador. – Publ. Field Mus. Nat. Hist., Bot. Ser. 7: 177–252.

– 1931b. The Rubiaceae of Bolivia. – Publ. Field Mus. Nat. Hist., Bot. Ser. 7: 253–340.

– 1931c. The Rubiaceae of Venezuela. – Publ. Field Mus. Nat. Hist., Bot. Ser. 7: 341–486.

Steyermark, J. A. 1964. Rubiaceae. – In: B. Maguire & J. J. Wurdack (eds.), The Botany of the Guayana Highland V. – Mem. New York Bot. Gard. 10(5): 186–278.

– 1965. Rubiaceae. – In: B. Maguire (ed.), The Botany of the Guayana Highland VI. – Mem. New York Bot. Gard. 12(3): 178–285.

– 1967. Rubiaceae. – In: B. Maguire (ed.), The Botany of the Guayana Highland VII. Mem. New York Bot. Gard. 17(1): 230– 436.

– 1972. Rubiaceae. – In: B. Maguire (ed.), The Botany of the Guayana Highland IX. Mem. New York Bot. Gard. 23: 227–832.

Verdcourt, B. 1958. Remarks on the classification of the Rubiaceae. – Bull. Jard. Bot. Etat (Bruxelles) 28: 209–281.

An overview of the genus *Pteris* (Pteridaceae) in Ecuador

Alba Luz Arbeláez,

Departemento de Biología, Universidad de Antioquia, Apart. Aéreo 1226, Medellin, Colombia

Botanical Institute, Aarhus University, Nordlandsvej 68, DK 8240 Risskov, Denmark

The genus *Pteris* belongs to the family Pteridaceae which include 35 genera with a world-wide distribution; 22 genera are present in the Americas, 14 of which are found in Ecuador with ca. 110 species.

The family has been divided into six tribes, five of which are present in Ecuador: Taenitideae, Cheilantheae, Ceratopterideae, Adianteae, and Pterideae. In Ecuador, the tribe Pterideae includes three genera: *Acrostichum* with two species distributed along coastal Ecuador, often associated with mangrove forests, at elevations from 0–1,000 m; *Neurocallis* with one species known only from the Pichincha Province; and *Pteris* with 22 species occurring from low to high elevations in primary and secondary forests throughout much of Ecuador.

Pteris exhibits the following characteristics: 1) sporangia marginal, 2) scales on the upper side of the costa, and 3) spores with a distinctive equatorial flange. Six species groups of *Pteris* have been recognized by Tryon and Tryon (1982), five of which are represented in Ecuador.

Two groups of *Pteris* may have escaped from cultivation and become established as naturalized species in Ecuador: the *longifolia*-group and the *cretica*-group, each represented by one species, *P. vittata* and *P. cretica,* respectively. Both are reported from Venezuela, Colombia and Peru.

Three groups are native to Ecuador. The *deflexa*-group contains 11 species in Ecuador: *P. altissima, P. coriacea, P. deflexa, P. gigantea, P. livida, P. muricata, P. podophylla, P. propingua, P. speciosa, P. transparens,* and *P. aff. tripartita.* This group is characterized by 1) lamina 2-pinnate to 5-pinnate – pinnatifid at the base, and 2) venation free or reticulate. The *quadriaurita*-group contains 6 species: *P. biaurita, P. horizontalis, P. pungens, P. quadriaurita, P. petiolulata,* and *P. fraseri*. This group is characterized by 1) basal pinnae forked, forming an inferior and stalked basal pinnule, and 2) venation free or reticulate. Finally, the *haenkeana*-group contains 3 species: *P. grandifolia, P. haenkeana,* y *P. splendens*. This group is characterized by 1) lamina 1-pinnate, or 2-pinnate at the base, or the basal pinnae forked, and 2) venation always reticulate.

A number of *Pteris* species exhibit a distinct and restricted pattern of distribution while others have a rather broad pattern of distribution. The coastal region of Ecuador contains five species: *P. biaurita, P. gigantea, P. propingua, P. pungens* , and *P. transparens,* associated with wet primary forests at low elevations. The eastern side of the Cordilleras contains two species found from premontane to midmontane forests: *P. deflexa* associated with secondary forests, and *P. speciosa* associated with primary forests. Many species are distributed within the Cordilleras. In the northern Cordilleras two species are found: *P. podophylla* and *P. quadriaurita*, both associated with secondary forests from low to mid elevations, although *P. podophylla* also occurs at high elevations. In the southern Cordilleras *P. haenkeana* and *P. splendens* are found, both associated with primary forest. Three species are found on both sides of the Cordilleras from low to mid elevations: *P. fraseri, P. horizontalis,* and *P. livida*. Two species exhibit a broad distribution in Ecuador associated with secondary forests from low to high elevations: *P. altissima* and *P. muricata*. Only one species, *P. coriacea*, is restricted to high elevations, often in subpáramo zones. Finally, *P. aff. tripartita* may be a new species known only from the Carchi Province at 1,000 m altitude, in wet primary forests.

References

Tryon, R. M. & Tryon, A. F. 1982. – Ferns and allied plants with special reference to tropical America. Springer, New York.

The genus *Acacia* (Mimosaceae) in Ecuador

Eva Bjerrum Madsen

Botanical Institute, Aarhus University, Nordlandsvej 68, DK-8240 Risskov, Denmark

Approximately 10 species of *Acacia* occur in Ecuador. They can be separated into two clearly distinct groups. The first consists of four species of shrubs or trees with bright yellow and often very fragrant flower-heads and armed with stipular spines. These species are found in the drier parts of the coastal lowland. One, *A. macracantha*, ascends into the dry valleys of the Andes, reaching an altitude of 2,200 meters above sea level. The second group consists mainly of climbers and shrubs, although one species, *A. glomerosa*, grows into big trees of 10–30 m height. The species in this group all have scattered and often recurved prickles and non-spinescent stipules. Their flowers are mostly white or pale yellow, gathered in axillary spikes or in distal panicles of small flower-heads. This group comprises at least five species, mostly found in the more humid forest types in the lowlands both east and west of the Andes. They are often growing in disturbed habitats, along rivers and roadsides, and in areas of tree-fall.

With a total of more than 1,200 known species, *Acacia* is the second largest genus in the legume family. During recent years it has gradually become more evident to researchers in various fields that this large genus must be at least biphyletic in its origin.

This has led to a suggestion from the Australian botanist Pedley, that the genus should be divided into three separate genera: *Acacia*, *Senegalia* and *Racosperma*. Most researchers agree that Pedley's new genus *Acacia (sensu stricto)* is easily separated from the remaining two groups. However, the distinction between the two latter genera is much more vague. Therefore most taxonomists consider Pedley's suggestion "premature".

One of the main reasons that taxonomists hesitate to split *Acacia (sensu latu)* into two or more genera is, that the New World species are very poorly known. Once these species become better explored it will be easier to judge whether Pedley's three genera constitute natural entities.

The genera *Brownea* and *Browneopsis* (Caesalpiniaceae) in Ecuador

Bente Bang Klitgaard

Botanical Institute, Aarhus University, Nordlandsvej 68, DK-8240 Risskov, Denmark

The genus *Brownea* was described by Baron N. J. Jacquin in 1760. Today the genus comprises 12 or13 species of trees and lanky shrubs ranging from Costa Rica to Peru, from sea level to 1,000 m. In Ecuador, *Brownea* is represented by four species.

In 1906 J. Huber described *Browneopsis*, a genus with three species, that is closely related to *Brownea*. This genus now comprises six species, with a western neotropical range from Panama to Peru, from 135 to 800 m above sea level. *Browneopsis* has three species represented in Ecuador.

There are still discussions as to whether *Browneopsis* should be recognized as a distinct genus or included in *Brownea*. Neither of the genera have ever been monographed, but several botanists have discussed the intergeneric relationship between them (Huber 1906, Pittier 1916, Macbride 1943, Velásques 1981, and Quiñones 1986). The latest treatments of *Brownea* are from Venezuela (Velásquez 1981) and from Colombia (Quiñones 1986). Both authors have chosen to keep them as distinct genera. My own morphological studies of the two genera revealed the following patterns.

Both genera have paripinnate leaves with a few to ca. 25 pairs of coriaceous leaflets. *Brownea* generally has more leaflets than *Browneopsis*.

Brownea has showy, red inflorescences borne in dense or loose capitulas, cauliflorous or terminal on branches, except in one species, *B.leucantha*, which has creamy white inflorescences. This species has been interpreted as a transitional form between *Brownea* and *Browneopsis* (Pittier 1916). The flowers are pedicillate, have large sheathing bracteoles, four equal sepals, five clawed, normally equal, red petals, and 11 or rarely 10 stamens. All species of *Brownea* are supposedly hummingbird pollinated.

The species of *Browneopsis* have by creamy white or rarely pale pink inflorescences, borne in compact capitulae, which are cauliflorous or terminal on branches. The flowers are sessile, ebracteolate, with 2–4 unequal sepals, 3–4 unequal petals, and 9–26 stamens. Morphological characteristics and field observations indicate that *Browneopsis* is bat pollinated.

Pollen morphological studies in Ecuadorean species have shown that *Brownea* has tricolporate, prolate, spheroidal or subprolate pollen, with a reticulate, striate to reticulate, or striate sculpture pattern. *Browneopsis* has tri- or tetraporate, oblate spheroidal pollen with a coarsely verrucate sculpture pattern.

In conclusion, both morphological and palynological studies in the Ecuadorean species of the two genera suggest that *Brownea* and *Browneopsis* should be kept as two distinct genera.

References

Huber, J. E. 1906. Materias para a Flora amazonica. – Bol. Mus. Paraense Hist. Nat. 4: 565–567.

Jacquin, N. J. 1760. Enumeratio Systematica Plantarum. – Leiden, pp. 6A, 26.

Macbride, J. F. 1943. Flora of Peru III(1). – Field. Mus. Nat. Hist. Bot. Ser. XIII: 131–135.

Pittier, H. 1916. The Genera *Brownea* and *Browneopsis* as represented in Panama, Colombia and Venezuela. Need of a new Treatment. – Contr. U.S. Nat. Herb. 18: 145–157.

Quiñones, L. M. M. 1986. Revision de las especies Colombianas del genero *Brownea* Jacq. (Leguminosae-Caesalpinioideae). – Universidad Nacional de Colombia, Bogotá.

Velásques, D. C. 1981. Revision taxonómica del género *Brownea* Jacq. (Leguminosae-Caesalpinioideae). – Universidad Central de Venezuela, Facultad de Ciencias, Escuela de Biología, Caracas.

Moraceae and Cecropiaceae in Ecuador

Cornelius C. Berg

Arboretum and Botanical Garden, University of Bergen, 5027 Bergen, Norway

Sixteen of the nineteen neotropical genera of Moraceae occurring in the Neotropics, occur in Ecuador. The number of species recorded in Ecuador till now is ca. 80. Further exploration will probably increase the number. The three neotropical genera of the Cecropiaceae are represented in Ecuador by ca. 50 species.

The taxonomy of most of the genera has been sorted out through revisional studies. A few minor problems remain to be solved in these, and only one or a few new taxa will have to be described. The two large genera, *Ficus* (125–150 spp.) and *Cecropia* (80–100 spp.), still have to be revised. The present state of the taxonomic knowledge of these two genera checks the preparation of the treatment of the two families for the Flora of Ecuador. The genus *Ficus* may not be as well represented as one would expect. Further exploration, focussing on tree species, may adjust the number of species.

Ten taxa have ranges of distribution that are probably almost confined to Ecuador. The genera *Cecropia* and *Ficus* may add a few taxa to this number of (sub) endemic taxa. The endemism is clearly connected with the Pacific lowlands and the Río Napo region.

A brief survey of the genera within Senecioneae (Asteraceae) in Ecuador

Roger Lundin

Museum of Natural History, P. O. Box 50 007, S-10405 Stockholm, Sweden

The Senecioneae (Asteraceae) are cosmopolitan and comprise approximately 100 genera and 2000 species. Traditionally the Senecioneae have been characterized by an epaleate receptacle and a pappus of capillary bristles. It is now more narrowly circumscribed and characterized mainly by an involucre with uniseriate or sometimes biseriate bracts with or without an outer calyculus of smaller bracts.

The Senecioneae are particularly distinct in their secondary compound chemistry. They accumulate pyrrolizidine alkaloids and sesquiterpenes of the furanoeremophilane type, whereas polyacetylenes are uncommon or absent.

The Senecioneae consist of two subtribes, Senecioninae and Blennospermatinae, according to Nordenstam (1978). Recently, Jeffrey (1984) divided Senecioneae into three subtribes: the Senecioninae, the Tussilagininae, and the Tephroseridinae. He excluded the subtribe Blennospermatineae which got a more separate position. With this classification, the Senecioneae could be regarded as consisting of four groups. The closest relatives are considered to be the Calendulae and/or parts of the Heliantheae and Anthemidae. The Senecioneae are one of the more specialized tribes within the subfamily Asteroideae (Bremer 1987).

There are two complexes of the subtribe Senecioninae (sensu Nordenstam) which are separated by a syndrome of morphological and cytological characters and which could very well be considered as subtribes as well.

The second subtribe, (sensu Nordenstam), Blennospermatinae comprises four small genera with obscure affinities *(Blennosperma, Crocidium, Ischnea,* and *Abrotanella).* None of these are present in Ecuador.

The Senecioninae are divided into two subgroups:

Cacalioid	**Senecioid**
Filament collar cylindrical basally	Filament collar swollen basally
Stigmatic surface entire	Stigmatic surface separated
Endothecial tissue polarized	Endothecial tissue radial
x=30	x=10
Heads often discoid and white-flowered	Heads radial and yellow-flowered
Involucre simple	Involucre with outer calyculus

The Senecioneae comprise about 12 genera with ca. 100 species in Ecuador, ranging from herbaceous plants in dry areas and widespread roadbank weeds to extremely adapted high-elevation herbs, shrubs and trees. They could be divided into the two subgroups:

Cacalioid group (Tussilagininae, sensu Jeffrey)
Gynoxys (ca. 25 spp.)
Aequatorium (4 spp.)
Senecioid group (Senecioninae)
Culcitium (4 spp.)
Aetheolaena (ca. 10 spp.)
Lasiocephalus (3 spp.)
Pseudogynoxys (4 spp.)
Pentacalia (ca. 24 spp.)
Dorobaea (2 spp.)
Senecio (c. 20 spp.)
Werneria (c. 10 spp)
Emilia (2–3 spp.)
Garcibarrigoa (1 sp.)

References

Bremer, K. 1989. Tribal interrelationship of the Asteraceae. – Cladistics 3(3): 210–253.
Jeffrey, C. 1984. Taxonomic studies on the tribe Senecioneae (Compositae) of Eastern Asia. – Kew Bull. 39(2): 205–454.
Nordenstam, B. 1978. Taxonomic studies in the tribe Senecioneae (Compositae). – Opera Bot. 44: 1–84.

Melastomataceae in Ecuador – what is new since 1980?

Susanne S. Renner

Botanical Institute, Aarhus University, Nordlandsvej 68, DK 8240 Risskov, Denmark

The Melastomataceae were treated by John Wurdack in volume 13 of the Flora of Ecuador, published in 1980. He dealt with 33 genera and 450 species - some of them only predicted to occur in Ecuador being known from near-by Peru or Colombia. The bulk of the material available to Wurdack had been collected in the 1960s, though he was able to include some collections from the late 1970s. The very large number of collections made in the last 10 years have resulted in a 22% (100 spp.) increase in the number of melastomes known from Ecuador. For species already known from the country, there are 125 new distributional records at the level of province. These figures are based on a survey of the material in AAU and US and on a comparison of notes with Dr. Wurdack in August 1989; included are the species collected by Missouri's David Neill and collaborators (most importantly Cerón, Palacios, and Zaruma) and identified by Wurdack.

At least 72 of the newly found species are named, some 30 - mostly in the large genus *Miconia* - remain undetermined. Most of these species are known from either Peru or Colombia. Of the named species, 51% come from Napo, 18% from Pastaza, and another 18% from Carchi which is floristically very close to Colombia. Of the new distributional records, 25% are from Napo, 18% from Pastaza, 13% from Carchi, and 10% and 9% from Pichincha and Esmeraldas, respectively. Carchi also harbours three of the six genera new to Ecuador: *Alloneuron* (*A. ecuadorense* Wurdack, Cyphostyleae), *Diplarpea* (*D. paleacea* Triana, Bertolonieae), and *Killipia* (*K. quadrangularis* Gleason, Miconieae). *Pilocosta* (*P. oerstedii* (Tr.) Almeda and Whiffin, *P. nana* (Standl.) A. and W., Tibouchineae) has been found in Pichincha and *Myrmidone* (*M. macrosperma* (Mart.) Mart., Miconieae) and *Tessmannianthus* (*T. cenepensis* Wurdack, *T. heterostemon* Mgf., Merianieae) in Napo and Pastaza. The nicest find perhaps is *Diplarpea*, the first collection of a genus only known from a single Colombian collection made by André in 1876. Since one genus recognized in 1980 has been sunk (*Platycentrum* = *Leandra*), Ecuador now has 38 genera, with ca. 550 species.

The province richest in melastomes is Napo with 170 named species, followed by Pastaza (106 spp.), Pichincha (66 spp.), Carchi (64 spp.), Morona-Santiago (60 spp.) and Esmeraldas (45 spp.). The eastern lowlands below 600 m altitude have 169 melastomes, the Andes above 2,400 m about 150, and the Andean slopes have at least 350.

Given the back-log of material awaiting processing I feel that there may well be as many as 600 species in Ecuador, *i.e.*, 50 more than presently known or a third more than treated in the Flora of Ecuador.

Palms of Ecuador

Henrik Balslev

Botanical Institute, Aarhus University, Nordlandsvej 68, DK 8240 Risskov, Denmark

This note reports on the present status of the project "Palms of Ecuador" which is carried out at the Botanical Institute of Aarhus University. The purpose of the project is to study Ecuadorean species of the palm family in order to 1) produce a complete inventory of Ecuadorean palms, 2) understand their taxonomy, 3) understand their biology, and 4) estimate their economic potential. The project has involved the collaboration of five masters level students, three doctoral students, and colleagues from USA, Colombia, and Ecuador.

Fieldwork was started with support from the Latinreco S. A. research laboratory during my residence in Quito from 1981 to 1984. Since 1984 it has continued with support from the Danish Natural Science Research Council. During a series of expeditions to Ecuador in 1985, 1986, and 1987 well over 1,000 collections of Ecuadorean palms have been made, together with many observations on their uses, their biology, and their economic potential in different land-use systems.

The project benefitted from the collaboration of several specialists. Gloria Galeano, Bogotá, participated in fieldwork in 1987 and now carries out monographic work on the genera *Ceroxylon, Prestoea,* and *Euterpe*. Rodrigo Bernal, Bogotá, participated in fieldwork in 1987. He is particularly interested in *Wettinia, Catoblastus,* and *Aiphanes,* and has contributed to the knowledge of these genera. Andrew Henderson, New York, has treated *Dictyocaryum, Iriartea,* and *Socratea* for Flora Neotropica and is now working with Gloria Galeano on a monograph of *Euterpe* and *Prestoea*. He has participated in fieldwork in Ecuador on several occasions.

The palm flora of Ecuador is rich and varied. Until recently it had only been studied sporadically, and the scientific literature contained only scattered references to records from Ecuador. Glassman's "A revision of Dahlgren´s Index of American Palms", published in 1972, listed 46 species of palms which had been recorded from Ecuador. At the end of our intensive fieldwork in 1987, the number of palm species known to occur in Ecuador had risen from 46 to 129 (Balslev and Barfod 1987). Since 1987 intensive studies of several genera have changed the numbers of species accepted in each of them; however, the total number remains the same. In addition we had records of 14 palm species which were cultivated in Ecuador. This number has increased by one to 15 species during the same period, an increase due mostly to the discovery of Ecuadorean populations of species already known from the neighboring countries, but several new species have also been discovered.

The Ecuadorean species of *Chamaedorea*, *Dictyocaryum*, *Iriartea*, *Socratea*, *Hyospathe*, *Attalea*, *Scheelea*, *Maximilliana*, *Aiphanes*, *Geonoma*, *Palandra*, *Phytelephas*, and *Ammandra*, have been treated in detail, and now include 58 species.

Ceroxylon, *Catoblastus*, *Wettinia*, *Euterpe*, and *Prestoea* are being worked on, and they are estimated to include 28 species in Ecuador

Some neotropical palm genera have been monographed recently. This is true for *Jessenia* , *Oenocarpus*, *Syagrus*, and *Parajubaea*. These four genera include seven species in Ecuador.

Finally, 12 genera and about 36 species of Ecuadorean palms have so far not been studied taxonomically in detail. The most difficult of these is *Bactris* with an estimated 15 species in Ecuador. *Desmoncus* has about five species in Ecuador and their naming is not easy because a great number of species have been described in that genus. *Astrocaryum* has four well defined species in Ecuador. The remaining untreated species belong to well known, mostly monotypic or small genera.

Pollination biology has been studied in *Phytelephas* and *Geonoma*. The ethnobotany of Ecuadorean palms has been studied among the Cayapas, Coaiqueres, Sionas, and Quichua indians. The economic potential of *Mauritia flexuosa*, *Astrocaryum jauari*, *Jessenia bataua*, *Ammandra natalia*, and *Euterpe chaunostachys* was the subject of a M. Sc. thesis by H. Borgtoft Pedersen, that of *Attalea colenda* was studied by U. Blicher-Mathiesen as part of her M. Sc. project.

References

Balsev, H. and Barfod, A. 1987. Ecuadorean palms – an overview. – Opera Bot. 92: 17–35.

TABLE 1. Ecuadorean Palm genera and numbers of species

SUBFAMILY Genus	No. of species 1972	 -87	 -89	 Collaborator
Native:				
CORYPHOIDEAE				
1. *Chelyocarpus*	1	1	1	
CALAMOIDEAE				
2. *Mauritia*	0	1	1	
3. *Mauritiella*	0	2	2	
CEROXYLOIDEAE				
4. *Ceroxylon*	2	4	8	G. Galeano Garcés
5. *Synechanthus*	1	1	1	Monogr. by H. E. Moore jr.
6. *Chamaedorea*	5	9	3	B. Bergmann
ARECOIDEAE				
7. *Dictyocaryum*	1	1	1	A. Henderson
8. *Iriartea*	1	1	1	A. Henderson
9. *Socratea*	2	4	4	A. Henderson
10. *Catoblastus*	0	2	4	R. Bernal
11. *Wettinia*	1	5	5	R. Bernal
12. *Manicaria*	0	1	1	
13. *Euterpe*	4	5	5	
14. *Prestoea*	0	6	6	
15. *Oenocarpus*	0	1	2	Monogr. by M. Balick
16. *Jessenia*	0	1	1	Monogr. by M. Balick
17. *Hyospathe*	3	5	2	F. Skov
18. *Cocos*	0	1	1	
19. *Syagrus*	1	3	3	Monogr. by S. F. Glassman
20. *Parajubaea*	0	1	1	Monogr. by A. Henderson and M. Moraes
21. *Attalea*	0	1	1	U. Blicher-Mathiesen
22. *Scheelea*	1	1	2	U. Blicher-Mathiesen
23. *Maximiliana*	0	1	1	U. Blicher-Mathiesen
24. *Elaeis*	0	1	1	
25. *Aiphanes*	3	5	13	F. Borchsenius
26. *Bactris*	2	15	15	
27. *Desmoncus*	0	5	5	
28. *Astrocaryum*	2	4	4	

TABLE 1. Continued...

SUBFAMILY Genus	No. of species 1972	 -87	 -89	 Collaborator
29. *Pholidostachys*	0	2	2	F. Skov
30. *Welfia*	0	1	1	F. Skov
31. *Geonoma*	15	34	27	F. Skov
PHYTELEPHANTOIDEAE				
32. *Palandra*	0	1	1	A. Barfod
33. *Phytelephas*	1	1	1	A. Barfod
34. *Ammandra*	0	2	2	A. Barfod
TOTAL	**46**	**129**	**129**	
Cultivated:				
CORYPHOIDEAE				
35. *Trachycarpus*	-	1	1	
36. *Chamaerops*	-	1	1	
37. *Livistona*	-	1	1	
38. *Pritchardia*	-	1	1	
39. *Washingtonia*	-	1	1	
40. *Sabal*	-	1	1	
41. *Phoenix*	-	3	3	
ARECOIDEAE				
42. *Chrysalidocarpus*	-	1	1	
43. *Roystonea*	-	1	1	
[*Archontophoenix*]	-	1	0	
44. *Veitchia*	-	0	1	
45. *Bentinckia*	-	0	1	
46. *Jubaea*	-	1	1	
47. *Elaeis*	-	1	1	
TOTAL	**-**	**14**	**15**	

Attalea colenda (Arecaceae), a potential lauric oil resource

Ulla Blicher-Mathiesen

Botanical Institute, Aarhus University, Nordlandsvej 68, DK 8240 Risskov, Denmark

Fruit production of *Attalea colenda* has been investigated. The palm is common on the coastal plain of Ecuador and in the southwestern part of Colombia. Its common name is "palma real" in northwest Ecuador and "chivila" in southwest Ecuador. The local people use the mesocarp and seeds of *A. colenda* fruits as a source of vegetable oil.

When the fruits are mature they fall to the ground where they are collected. In former times the sweet and oily mesocarp was consumed directly. The mesocarp becomes loose after boiling, and the oil can be collected from the surface of the boiled mass. Today, individual trees are jealously guarded by small holders, who exploit the seeds. The hard endocarp is crushed with a hammer or a stone in order to release the seeds. These are then used to feed pigs or are sold to agricultural warehouses in nearby towns.

About four kilometers from the town of Manta, along the road to Montecristi, the factory "Ales" extracts oil from the seeds. According to an engineer at the factory, water content of the seeds is 20–25% when they arrive. They are therefore dried before the oil is extracted. The oil content of the seeds is between 50–58%. The oil is used for consumption and soap. In 1987, farmers in the province of Esmeraldas said that the labor involved in releasing the seeds with a hammer made the exploitation of the palm uneconomic. Apparently, no machine has yet been developed to facilitate the extraction of the seeds.

In a one-hectare plot in a pasture near Borbón, fruit production was estimated. Thirteen infructescences of *A. colenda* were harvested from eight individuals and the number of rachillae per infructescence counted. In a line from the base to the apex, 30 rachillae were selected and the number of fruits per rachilla counted. The fruit number per infructescence was 3,500–7,100 with a mean of 5,065. The fruit weight relative to infructescence weight was on average

86% (SD = 3.8; min. = 81%; max. = 95.5%; n = 17). A sample of 50 fruits contained 49% mesocarp, 32% endocarp, and 19% endosperm (kernel) of the fresh weight. Mean weight of 100 dried kernels was 3.94 grams. Calculated oil production per infructescence was 7–16 kg (mean = 10.6 kg). In Borbón, in 1988, the first fruits were mature in the first week of January. At this time, 25 individuals did not produce fruits, eight palms had one infructescence, six palms had two, and two palms had three infructescences. In one palm, the apical meristem was probably dead. Thus, in 1988, the oil production of 42 palms investigated in the pasture hectare would have been approximately 276 kg. *Attalea colenda* starts to produce 1–2 infructescences per year at an age of 10–12 years. When it is mature it produces 3–4 infructescences per year. The largest number observed was seven. With three to four infructescences per palm, it can be calculated that 50 mature trees can produce between 0.35 and 3.2 tons of oil per year per hectare.

The kernel oil of *Attalea colenda* is similar to the kernel oil of the African oil palm (PKO) and coconut oil. These oils have relatively low melting points, are hard at lower temperatures, and when saturated they are very resistant to oxidation. The kernel oil of of *A. colenda* comprises 47.2 % lauric acid. Lauric acid is used for soap, dairy fat replacement in margarine, biscuits, imitated cream, and filled milk.

Considering that *A. colenda* is a wild species in which no artificial genetic improvement has taken place, an oil production of 0.35–3.2 tons per hectare makes it extremely interesting.

Phytogeographical patterns of Ecuadorean grasses

Simon Lægaard

Botanical Institute, Aarhus University, Nordlandsvej 68, DK 8240 Risskov, Denmark

In 1986, a provisional check-list of Ecuadorean grasses was prepared, including records from Hitchcock (1927), Acosta-Solís (1969) and several others. A total of 453 species were recorded, but since then at least 15–20 new species have been added from the limited part of the family that has been studied. It is cautiously estimated that the total number will be more than 500.

Subfamily	genera	species
Bambusoideae	16	35
Pooideae	31	134
Centothecoideae	2	2
Arundinoideae	6	10
Chloridoideae	21	77
Panicoideae	39	195
Total	**115**	**453**

Each of the subfamilies has a distinct phytogeographic profile.

Panicoids are almost exclusively in the tropical lowlands up to about 2,000 m, only a single species, *Paspalum bonplandianum,* belongs to the páramo vegetation at 4,000 m or higher. Among the Panicoids are a number of pantropical weedy plants - for several of these the original homeland is hardly known. They are usually found in open, disturbed sites as, *e.g.*, cultivated fields and roadsides. The native species are found in shaded as well as in open places, but often in moist biotopes, such as riversides and swamps.

Economically important cultivated species are *Zea maiz* and *Saccharum officinale. Axonopus scoparius* (a native species), *Panicum maximum* (pantropical) and *Pennisetum purpureum* (African) are important pasture plants in the tropical lowland.

Pooids, on the other hand, are nearly all found above 2,000 m, only the two species of *Polypogon* are regularly found in tropical lowlands. Most grass species of the páramos belong to the Pooids and most of the genera occurring are also common in the northern temperate zone, *e. g.*, *Calamagrostis, Festuca, Agrostis, Poa,* and *Bromus*. Some of these genera are very diverse with many species that are well adapted to the burning of páramos, and some are only found in regularly burned areas.

Several introduced and cultivated Pooids are important pasture-plants in the highlands, *e. g.*, *Dactylis* and *Lolium*. *Poa annua, Holcus lanatus, Anthoxantum odoratum* and others have been introduced and grow as weeds.

Chloridoids are found more or less evenly distributed from coastal salt marshes and riverbanks in the tropical lowlands to the highest páramos. Generally they are light demanding, most of them belong to dry biotopes but a few are found in wet places. They are rarely dominant in any vegetation type, except maybe as weeds.

The most prominent members of the subfamily Arundinoideae are *Cortaderia* of the páramos and *Gynerium* of the tropical lowlands. Among the five species of *Cortaderia* there is much variation in sexuality as they comprise both hermaphroditic, gynodioecious, dioecious, and apomictic species. The apomictic species, *Cortaderia jubata*, shows some variation and may be a future field for description of microspecies. *Gynerium sagittatum* is probably the largest of all the true grasses, bamboos excluded; in slightly protected swamps in the rain forest it can be 12–15 m high. *Arundo donax* has been introduced from the Old World; it is widely cultivated and escaped.

Bambosoideae are usually treated as three different groups: The real, woody bamboos, the herbaceous bamboos, and the rice-group.

The most important woody bamboo genus is *Chusquea* from the mountain forest regions and the páramos. It is ecologically important in many mountain regions, often it becomes weedy after forest clearance, and is an obstacle both to land-use and to reestablishment of the forest. In all lowlands, *Guadua* is of utmost economical importance as building material etc.

The herbaceous bamboos, *e. g.*, *Pharus* and *Pariana,* are almost exclusively found in tropical rain forests, most of them are very susceptible to disturbance and they are severely threatened by forest clearence.

In the rice-group there are, besides the cultivated rice, a local species of *Oryza* and a few other swamp plants of minor importance.

References

Acosta Solís, M. 1969. Glumifloras del Ecuador. – Inst. Ecuator. de Ciencias Nat., Contr. 71.

Hitchcock, A. S. 1927. The Grasses of Ecuador, Peru, and Bolivia. – Contr. US Nat. Herb. 24(8).

Intraspecific variation in *Kohleria* (Gesneriaceae)

Lars Peter Kvist

Botanical Institute, Aarhus University, Nordlandsvej 68, DK 8240 Risskov, Denmark

The genus *Kohleria* (Gesneriaceae; Gesneriodeae; Gloxinieae) has been revised. Species of *Kohleria* are attractive, terrestrial herbs usually with conspicuous red flowers. *Kohleria* ranges from Mexico to Peru and east to Suriname and from sea level up to an altitude of about 2,500 m; most commonly the species are found between 800 and 2,000 m.

During the last century *Kohleria* was popular in European greenhouses, and many species were described from cultivated material of unknown origin. For this reason the taxonomy of *Kohleria* has been highly problematic. The revisionary work demonstrated that although approximately 100 species have been described only 17 species can be recognized, and four of these are new. The diversity center is Colombia with a total of 14 species, nine of which are endemics.

Eleven *Kohleria* species occur mainly in exposed, disturbed habitats along rivers or roads, etc. They have capsules which split apically into two valves, and their tiny seeds are wind dispersed. In contrast, the remaining species occur in shaded, permanently humid understorey often close to streams. Wind-dispersal is no option in this environment. In these species the capsules split by a single slit from the apex to the base and expose a sticky seed mass which is apparently removed by unidentified nocturnal animals. The understorey species also lack the scaly rhizomes which enable species from exposed habitats to survive dry spells.

The three most common species, *Kohleria hirsuta*, *K. spicata*, and *K. tubiflora*, are all wind dispersed. The latter two species have surprisingly stable characteristics throughout their wide ranges, *e. g.*, *K. spicata* specimens from Mexico look very similar to those from Ecuador. In contrast, *Kohleria hirsuta* is extremely variable especially within the Colombian diversity center of the genus. This is partly due to hybridization with other species, *e. g.*, with *K. trianae* on the eastern slopes of the central Colombian cordillera. *Kohleria hirsuta* and *K. trianae* actually have allopatric ranges in the eastern and the central cordilleras, respectively. However, especially *K. hirsuta* is wind-dispersed across the dry, hot Magdalena river valley and hybridizes with *K. trianae*. The presence of the Magdalena valley apparently prevents hybridizations from completely breaking down the limits between these two species.

The most common and widespread understorey species, *K. inaequalis*, may be the most variable species of Gesneriaceae so far studied; variable features are the sizes of calyx lobes and corollas, the lengths of peduncles, and the lengths of trichomes. *Kohleria inaequalis* appears to be a complex of three semi-species that are hybridizing extensively since, at least at the present time, no dispersal barriers exist between their ranges. Pure-breeding populations of the three semi-species have peripheral ranges and are characterized by having either large flowers and pedunculate inflorescences or large flowers and epedunculate inflorescences or else small flowers and pedunculate inflorescences.

Generally, the predominant mode of speciation in *Kohleria* appears to be geographical isolation of peripheral populations, while hybridization is breaking down the limits of species rather than creating new species.

The diversity of micromorphological features in *Coccoloba* (Polygonaceae)

John Brandbyge

Botanical Institute, Aarhus University, Nordlandsvej 68, DK 8240 Risskov, Denmark

Coccoloba P. Browne ex L. is a large and taxonomically complex genus of woody Neotropical Polygonaceae. In this study, which includes 70 taxa, leaf and perianth epidermis and pollen surfaces were studied by scanning electron microscopy (SEM). Stomata, primary sculpture, epicuticular wax deposits, and multicellular glandular hairs of leaf blades together with secondary sculpture of perianth epidermis constitute taxonomically important micromorphological characters. Pollen morphology of the examined species is relatively uniform. Types based on surface ornamentation vary continuously from finely punctate-striate to deeply punctate-striate or microreticulate. Two species, *C. acapulcencis* and *C. cordata,* differ markedly from the rest of the species investigated by having spinulose ektexines. The diagnostic-descriptive value of the characters studied is more pronounced than the taxonomic-phylogenetic value. The two closely related species pairs *C. acapulcencis* /*C. cordata* and *C. brasiliensis* / *C. schomburgkii* have disjunct distribution.

The biogeography of *Oreobolus* (Cyperaceae)

Ole Seberg

Botanical Laboratory, University of Copenhagen, Gothersgade 140, DK 1123 Copenhagen K, Denmark

The genus *Oreobolus* R. Br. (Cyperaceae), with four taxa in the northern Andes, is well suited for biogeographical analysis (Seberg 1988). It is beyond reasonable doubt monophyletic, a phylogenetic hypothesis, based on the principle of parsimony, is available, and its 17 taxa (14 species, three of them with two subspecies) have an Austral-Antarctic distribution pattern, ranging from the Flora Malesiana-area across southeastern Australia, Tasmania, New Zealand, Hawaii, Tahiti, and up along the Andes from Tierra del Fuego in the south to Venezuela in the north, with a distinct distribution gap between Mendoza and northernmost Peru. Additionally it is known from Central America (Costa Rica, Panama) and from the Serra de la Neblina, Brazil.

The four north Andean/Central American species of *Oreobolus*, *O. ecuadorensis*, *O. venezuelensis*, *O. goeppingeri*, and *O. obtusangulus* ssp. *unispicus*, are endemic to the páramo and thus have an island-like distribution pattern. This patchy distribution pattern immediately raises two important biogeographic questions: 1) What are the area-relationships indicated by these páramo-taxa to areas elsewhere in the distribution range?, and 2) What are the mutual area-relationsships indicated between the individual groups of páramo "islands" ?

With respect to the first question, the available area cladograms for other taxa with Austral-Antarctic distribution pattern (however, not necessarily in details sharing the same overall distribution pattern), viz. *Nothofagus* (Fagaceae), *Cyttaria* (Helotiales), and the Eriococcidae (Homoptera), which are mutually congruent, are incongruent with the one obtained for *Oreobolus*. In terms of vicariance the two resulting, conflicting area cladograms are equally parsimoniously explained by a wide variety of historical-geological hypotheses describing the break-up sequence of Gondwanaland, even with some incorporating the hypothetical continent of Pacifica.

Focusing alone on the northern South American species of *Oreobolus,* no other vicariance cladistic hypotheses about the area relationship of the páramo "islands", than the one for *Oreobolus*, are found. It is thus impossible to give any indication about the generality of this pattern. Due to the very extensive sympatry of the four taxa little information about the relationships of the individual paramo "islands" may be obtained. The most noteworthy features of the area cladogram based upon *Oreobolus* are the facts that imbeded in the hypothesis, is a relationship to Australia-Tasmania and a northern-southern South American relationship. The latter disjunction is not uncommon, *e.g.*, it is also found in *Rostkovia* (Juncaceae), *Lagenophora* (Asteraceae), *Gunnera* (Gunneraceae). The merit of the former area relationship may, however, only be judged in the light of further investigations.

References

Seberg, O. 1988. Taxonomy, Phylogeny, and biogeography of the genus *Oreobolus* R. Br. (Cyperaceae), with comments on the biogeography of the South Pacific continents. – Bot. J. Linn. Soc. 96: 119–195.

Floristic composition, altitudinal distribution, and biogeograpical patterns of the Colombian moss flora

Steven P. Churchill

New York Botanical Garden, Bronx, NY 10458, USA

–

Botanical Institute, Aarhus University- Nordlandsvej 68, DK 8240 Risskov, Denmark

The Colombian moss flora promises to be one of the richest of South America given the varied geological history, existing topography, and the wide range of vegetation types. A project has recently been initiated to conduct a systematic field survey and produce a moss flora for Colombia. Past field work by the author has included surveys of the departments of Antioquia and Valle, with preliminary work in the departments of Caldas, Chocó and Nariño.

Support has previously been received from the New York Botanical Garden in cooperation with the University of Antioquia, and recently from the Danish Research Academy. Future field work in the coastal and western Cordilleras of Colombia and preparation of the first half of the manuscript (covering the pleurocarpous mosses) will be supported by the National Science Foundation, USA. An additional three years will be required to complete field work in the Llanos and Amazonas of eastern Colombia, and portions of the central and eastern Cordilleras of Colombia, as well as the completion of the manuscript (acrocarpous mosses *etc.*).

An estimate of the floristic composition can be derived from a recently completed revised checklist for Colombian mosses that I have prepared (1989, Trop. Bryol. 1: in press). A total of 877 species distributed among 242 genera and 65 families are presently recorded. The ten largest families are: Dicranaceae (110 spp.), Callicostaceae (71 spp.), Bryaceae (67 spp.), Pottiaceae (63 spp.), Orthotrichaceae (55 spp.), Bartramiaceae (51 spp.), Sematophyllaceae (44 spp.), Brachytheciaceae (31 spp.), Fissidentaceae (30 spp.), and Polytrichaceae (29 spp.). These ten families account for ca. 62% of the total moss flora of Colombia.

The actual number of species present in Colombia is still uncertain. Since the completion of the first checklist by Florschütz-de Waard and Florschütz in 1979 (Bryol. 82: 215–259) who listed 750 species, ca. 140 new records, including seven new species, have been added to the Colombian moss flora. The revised checklist for Colombia contained 40 additional species as well as 560 new departmental records for Colombia. These data stress the need for further systematic field work. More than 75% of the 32 departments in Colombia are poorly known. Departments with mostly montane and páramo vegetation have received the greatest attention in the past. The lowland tropical areas, both Llanos and wet and dry tropical forests, have received no serious field work. Of 12 departments 10 or less species recorded, 11 are from the lowlands where 50 to 100 species should be expected in each. A serious effort at a systematic field survey of Colombia, could provide a better knowledge of distributional patterns, as well as increase the number of known species. Systematic studies, however should reduce the number of names in the present checklist. Given both factors, further field work, and systematic studies, the actual number will probably lie close to 800 species.

Analysis of altitudinal distribution of Colombian mosses was based on data available for ca. 750 species. Information was derived from existing literature, herbarium collections, particularly in NY and U, and personal field collections. Six zones were employed, modified from those used previously in Colombia by van Reenen and Gradstein in their ecological bryophyte study (1984, Jour. Hattori Bot. Lab. 56: 79–84). The following results give the altitudinal zone and the species number recorded for each: 1) 0–600 m, 103 spp.; 2) 600–1,400 m, 144 spp.; 3) 1,400–2,000 m, 218 spp.; 4) 2,000–2,600 m, 329 spp.; 5) 2,600–3,300 m, 382 spp.; 6) 3,300 + m, 260 spp. Many species exhibit a rather wide elevational amplitude, occuring in two or more zones. Pleurocarpous mosses were dominant in zones 1–4, whereas the acrocarpous mosses were dominant in zones 5–6.

The biogeographical patterns of the Colombian mosses are typical for most northern Andean countries, exhibiting recognizable and repeating distributional patterns. Two major components are recognized, Neotropical and Extra-Neotropical.

Naturally most species are associated with a Neotropical pattern, and of these most are rather widespread with a distribution extending beyond the Andes. Among these, mostly at mid- or high elevations, are species that are found in Mesoamerica, followed by those that are found as disjunctions with southern Brazil and less so with the West Indies. A number of widespread Neotropical species are found in the wet lowland forests. Those species primarily associated with the Andes tend to be somewhat more widespread within the Andean system, extending from Costa Rica south to Bolivia. A smaller number are strictly confined to the northern Andes of Venezuela, Colombia and Ecuador. Finally, possibly 100 or more species are recorded as Colombian "endemics." However, it is believed that this number is greatly inflated, and that many will prove to be synonyms of more widespread taxa. Affinities of Extra-Neotropical species lie mostly with so-called temperate, mostly northern, taxa. Such taxa are assumed, *a priori*, to have migrated and/or been dispersed from the north or south, and in Colombia they tend to occupy high elevations in the upper montane and páramo zones.

Finally, a small element is represented by pantropical species, mostly of African affinity, and to a lesser extent of Australasian.

Herbaceous ground flora in a tropical rain forest

Axel Dalberg Poulsen

Botanical Institute, Aarhus University, Nordlandsvej 68, DK 8240 Risskov, Denmark

An inventory was conducted in the Amazon lowland of Ecuador with the purpose of estimating the total number of species. The study was carried out in a one hectare sample plot of homogeneous *terra firme* rain forest in Reserva Faunística Cuyabeno in the Napo Province.

The herbaceous ground flora registered includes 95 species from 14 families. They are all perennials and they represent the following life forms:

1) *True terrestrials* (46 spp.) that are not dependent on physical support from other plants and complete their life cycle without loosing ground contact.
2) *Scandent climbers* (6 spp.) that depend to some extent on other plants for support, but otherwise persist having ground contact.
3) *Climbers* (23 spp.) that climb in trees or shrubs and only become fertile several meters above the ground, often after having lost their primary ground contact. The group is dominated by *Araceae*, some of these commenly have their juvenile stage on the ground.
4) *Epiphytes* (19 spp.) that have fallen down and become established on the ground and that may at least for some time survive on the ground.
5) *Saprophytes* (1 sp.),*Voyria flavescens,* was collected in the sample plot.

For each species the following was noted: 1) coordinates within the study plot, 2) number of individuals, 3) maximum height, 4) coverage, and 5) phenology

About 60% of all species were found in some fertile stage during the 10 weeks of fieldwork. Based on field notes the Importance Value Index (acc. to Curtis and MacIntosh 1951) for each species and the Family Importance Value (acc. to Mori *et. al.*1983) were calculated.

The most important group occuring in the sample plot was the ferns, followed by aroids and grasses. The most important species was *Pariana radiciflora* (Poaceae).

The total coverage for all ground herbs was estimated to be 2.5 %. Even though this may seem low, the diversity of the grund herbs is rather high.

The species/area curve for ground herbs clearly shows that a sample area of one hectare is sufficient to register most of the species of ground herbs occurring in this rain forest type, while it is known that it is usually too small for an inventory of the trees.

Maps of spatial distribution of each species fall into different categories: even, clustered, rare, and mega-clustered, in correllation with the topography and the affinity of the species to "wet parts" and "very wet parts". An understanding of these distribution patterns would be facilitated by information about seed production and seed dispersal syndrome of the individual species, as well as by more knowledge of the abiotic factors. Some of the distribution patterns clearly show a correlation with the topography and thereby probably with the water content of the soil.

References

Curtis, J. T. and McIntosh, R. P. 1951. An upland forest continuum in the prairie-forest border region of Wisconsin. – Ecology 32: 476–496.

Mori, S. A., Boom, B. M., Carralino, A. M. de, and Santos, T. S. dos 1983. Ecological importance of Myrtaceae in an Eastern Brazilian wet forest. – Biotropica 15: 68–70.

Epiphytes in a *terra firme* rain forest

Ingvar Nielsen

Botanical Institute, Aarhus University, Nordlandsvej 68, DK 8240 Risskov, Denmark

The purpose of this study was to register all species of vascular epiphytes within a sample plot of one hectare of tropical lowland rain forest in Reserva Faunística Cuyabeno, Napo Province, Ecuador. The herbaceous ground flora in the same hectare plot was studied by Axel Dalberg Poulsen (see previous contribution).

For each collection, notes were made on the ecology of the species, in order to get a more systematic knowledge and understanding of the ecology of epiphytes.

The broad definition of epiphytes suggested by Madison (1977) was adopted, which includes "all plants which at some stage in their life cycle are normally not connected to the ground by a stem".

For a more specific understanding of the epiphytic lifeform, the more narrow definition of Madison was found useful, saying that true epiphytes are "those species which normally germinate on the surface of another living plant and pass the entire life cycle without becoming connected to the ground".

In praxis, all trees, shrubs, and lianas in the sample plot were searched from the base and upwards, and the epiphytes were registered according to Johanssons (1974) division of the phorophyte into five zones. In this way, practically all species in zone 1 (up to 3 m height), some in zone 2, but very few from the canopy (zone 3–5) were recorded, since neither time, nor the equipment to reach the canopy were available.

A total of 131 species of epiphytes were found, according to the broad and 87 species according to the narrow definition. Of the 87 true epiphytes, 48 species were from zone 1, including 13 Pteridophytes, 13 Bromeliaceae, 9 Araceae, 5 Gesneriaceae, 2 Melastomataceae, 2 Piperaceae, 1 Orchidaceae, 1 Begoniaceae, and 1 yet unidentified.

Ten different parameters were used to characterize the ecology of each species of epiphytes. Among these were: 1) type of epiphyte, 2) substrate, 3) amount of light, 4) presence of ants, and 5) habitat. The work has not been completed yet, so only a few results concerning the first three parameters are presented here.

Of the 48 species in zone 1, 52 % were proto-epiphytes, 25 % tank- epiphytes, 8 % nest- and bracket-epiphytes, and 15 % both proto-epiphytes and nest- and bracket-epiphytes (definitions of types according to Schimper 1903). In zone 1, 81 % of the species grow on bark, 6 % grow in minor humus deposits, and about 13 % were both found in humus deposits and directly on bark. Finally, 85 % of the same species grow in open shade (mixture of shade and sun), 4 % in heavy shade (no direct sunlight), and 4 % in both types of light conditions.

References

Johansson, D. 1974. Ecology of vascular epiphytes in West African rain forests. – Acta Phytogeogr. Suec. 59.

Madison, M. 1977. Vascular epiphytes: their systematic occurrence and salient features. – Selbyana 2: 1–13

Schimper, A. F. W. 1903. Plant-geography upon a physiological basis. – Oxford

Structure and composition of two montane rain forests.

Peter Møller Jørgensen

Botanical Institute, Aarhus University, Nordlandsvej 68, DK 8240 Risskov, Denmark

The ecology of the montane rain forest is practically unknown. Due to the needs of firewood in the fast growing rural population of the Andes this vegetation type is disappearing with an alarming velocity. Two 1-hectare quadrats were established: 1) on Volcán Pasochoa, ca. 30 km southwest of Quito at an altitude of 3,300 meters; 2) in the Valley of Lloa ca. 22 km east of Quito at an altitude of 2,900 meters. The method applied included permanent marking of the quadrat and its trees; the quadrat was marked with PVC–tubes and all trees with a diameter larger than or equal to 5 cm at breast height (130 cm) were marked with aluminum tags. For each tree the following data were registered: position inside the quadrat, diameter, height, type and color of bark, bark thickness, color of the wood, and presence of latex. Notes were made on the presence of epiphytes and climbers. In the case of Pasochoa, a detailed description of the understorey was also compiled.

The results from Pasochoa can be summarized in 10 points:

1) 34 species of trees were found with a DBH ≥ 5 cm. The species belong to 31 genera in 21 families.
2) The minimum area, which includes 90 % of the species, was 8,500 m^2.
3) The density is 1,060 trees per hectare with a DBH ≥ 5 cm and 710 trees per hectare with a DBH ≥ 10 cm.
4) *Miconia theaezans* is the species with highest value both in the "Importance Value Index" and in the components of this index (relative values of dominance, frequency, and density).
5) The Basal Area (dominance) is concentrated in two species, *Miconia theaezans* and *Piper andreanum*, that amount to more than 50 % of the total basal area of 25.8 m^2.

6) The "Family Importance Index" shows that Melastomataceae is the ecologically most important family; furthermore this has the highest values of each component of the index (relative density, diversity, and dominance). The second-most important family is Piperaceae. The third most important family, Asteraceae, is, however, more diverse than Piperaceae.
7) The distribution of trees in diameter classes shows that the forest has been disturbed several years ago. Apparently, it has been selectively cut.
8) Epiphytes are very abundant, and the diversity is high in groups such as Orchidaceae, Bromeliaceae, mosses, and ferns. Epiphytes were found on 98 % of the trees, climbers on 79 %.
9) The understorey is dominated by mosses that cover ca. 38 % of the area, followed by *Chusquea scandens*, which covers ca.. 30 %.
10) Comparing the distribution of all the trees and individual species in diameter classes it is possible to detect whether or not changes have occurred in the forest. Changes in this apparently primary forest were detected, and later confirmed by peasants from the area, who told that the area was selectively cut 28–30 years ago.

The results from the quadrat in Lloa have so far not been fully analysed, and it is therefore premature to make a full comparison of the two quadrats. However, a few results can be given:

1) 785 trees were found with a DBH $\geq$ 5 cm, the density is thus considerably lower than at Pasochoa.
2) The three most important families are apparently Actinidiaceae, Meliaceae, and Euphorbiaceae.
3) Very few species known from Pasochoa have also been found in Lloa, however, many of the genera and 2/3 of the families are common to the two sites.
4) There seems to be a lack of young trees (diameter classes from 5–15 cm). This may be due to selective cutting of the smaller trees in order to make room for cattle grazing inside the forest. No larger trees have, however, been touched and the crown layer seems to be intact.

Trees and shrubs of the high Andes of Ecuador

Carmen Ulloa U.

Dept. de Ciencias Biologicas, Pontificia Universidad Católica, Apart. 2184, Quito, Ecuador

–

Botaniska Institutionen, University of Göteborg, Carl Skottsbergs Gata 22, S-413 19 Göteborg, Sweden

The high Andean forests of Ecuador have previously not been studied in detail. Their species composition, diversity, and natural resources are therefore practically unknown. Furthermore, people have been interacting with the vegetation for up to 10,000 years. Deforestation, cultivation, burning, grazing and the introduction of exotic species, such as *Eucalyptus* have completely changed the landscape.

The study "Genera of trees and shrubs of the high Andes of Ecuador" was initiated in order to get more knowledge about these forests. It includes descriptions of all families and genera, notes on their distribution, common names and uses, and a checklist of all native species found above 2,400 m. Upwards, trees and shrubs are found until 4,000 m or even higher.

The study is based on information from labels of herbarium specimens, records in literature, dissection of selected specimens and fieldwork. All Ecuadorian material collected above 2,400 m, deposited in Quito (QCA), Aarhus (AAU), Göteborg (GB), and Stockholm (S) and for special groups also in C, G, LD, N, and US has been registered.

The upper Andean forests are generally very humid and characterized by trees up to 15 m, often with twisted trunks covered by mosses. They are often dominated by one or a few species of the genera *Miconia, Saurauia*, and *Polylepis*. The most diverse families are: Asteraceae (ca. 34 genera), Ericaceae (18), Solanaceae (14), Melastomataceae (12) and Rubiaceae (9). Some of the largest genera are: *Miconia* (91 spp.), *Centropogon* (32 spp.), *Berberis* (31 spp.), *Calceolaria* (27 spp.), *Fuchsia* (23 spp.), *Baccharis* (21 spp.), *Brachyotum* (20 spp.), *Palicourea* (18 spp.), *Diplostephium* (17 spp.) *Monnina* (16 spp.) and *Gaultheria* (15 spp.).

At present 1,073 species of woody plants have been recorded, belonging to 239 genera and 73 families.

The Danish-Ecuadorian project in the Parque Nacional Podocarpus, Ecuador

Benjamin Øllgaard and Jens Elgaard Madsen

Botanical Institute, Aarhus University, Nordlandsvej 68, DK 8240 Risskov, Denmark

Parque Nacional Podocarpus (ca. 146,000 ha, provinces Loja and Zamora-Chinchipe, southern Ecuador) is situated across the Eastern Cordillera at 1,500–3,500 m. Most of it is covered by wet montane forest. Smaller areas above the forest limit (alt. 3,000 m) are covered by shrubby and herbaceous vegetation, corresponding to páramo or jalca vegetation, but until recently very rarely or not influenced by fire. The majority of the forest is virgin, but road construction, concessions for mining industry, illegal logging, and fire-raising contribute to diminish the virgin area.

The project in progress aims to produce a floristic survey, a study of forest structure and composition, and to find native trees for reforestation.

Two one-hectare plots have been established in undisturbed forest at 2,900 m in the Cajanuma area near Loja, and at 2,700 m in the Quebrada Honda, just south of Nudo de Sabanilla. Trees with DBH ≥ 5 cm were permanently tagged, mapped, measured, and vouchered. Their phenology is under observation. The preliminary results show that these plots differ in structure and composition from comparable plots in northern Ecuador. Tree number per ha is app. 2,500 in the first, and app. 2,200 in the second plot. The number of species per ha is ca. 60 in the first, and ca. 75 in the second plot. This is ca. twice as many trees per ha and species as recorded for corresponding plots in northern Ecuador (see P. M. Jørgensen, this volume). Dominant trees are species of *Weinmannia, Ternstroemia, Clusia, Ocotea, Persea,* Melastomataceae, Myrsinaceae, and Celastraceae in the first, and *Hedyosmum, Clusia,* Melastomataceae, *Podocarpus, Ceroxylon,* and *Saurauia* in the second plot.

In collaboration with Ministério de Agricultura y Ganadería, Universidad Nacional de Loja, and PREDESUR, a selection of native tree species from the area are artificially propagated in local plant nurseries. Their qualities and requirements are studied. The results will be used to increase the number of native trees for local reforestation programs. Special attention is given to reneval of endangered timber trees, that are especially valued by the local population, and to tree species for reforestation of degraded grassland in the drought-plagued valleys south of Loja.

Species of the following genera are currently tested: *Clusia, Hedyosmum, Laplacea, Myrcianthes, Myrsine, Ocotea, Persea, Podocarpus, Symplocos, Ternstroemia, Tibouchina, Trichilia,* and *Weinmannia.*

Regeneration strategies in gaps in a montane rain forest in southern Ecuador

Lars Christensen

Botanical Institute, Aarhus University, Nordlandsvej 68, DK 8240 Risskov, Denmark

The aim of the study was to describe the importance of and some of the interactions between different groups of plants occupying various regeneration niches in disturbed areas.

One study site (A) was placed in an area that had been clear-cut seven years ago, and in which the trunks had not been removed. Fifty sample plots of one square meter were established, and in each plot the following data were recorded for each species: 1) regeneration strategy type, 2) no. of individuals, 3) phenology, 4) coverage, and 5) maximum height. From this were calculated 1) frequency, 2) density and 3) number of species.

In another area (site B) sample plots of one square meter were placed. In this area all roots, stumps, and rhizomes had been removed seven years ago. Therefore the regeneration was entirely dependent on colonization and establishment of seedlings and perhaps also on a former soil seed bank. The soil was a mixture of the upper two meters of soil from the surrounding area, pushed together by a bulldozer.

The plant species were devided into groups according to niche affinity and regeneration strategy. These groups were: ferns, graminoids, epiphytes, forbs (herbs other than ferns, graminoids and epiphytes), tree ferns, palms, vines, shrubs, tree seedlings, and tree sprouts.

There are numerous interactions and overlaps, which determine the micro-distribution of the niches. The first and perhaps most conspicious phenomenon is the role of the bambusoid genus *Chusquea* in gaps of various types. At site A it covers the other vegetation totally. This cover favours the shade tolerant species throughout the first years of regeneration. At sites where *Chusquea* is not present, there is a greater dominance of seedlings and pioneers or shade intolerant species.

Because of their vigorous growth the bamboos cover the ground very soon after a clearing. This may prevent extensive soil erosion. At site B, where bamboos are absent, the clay fraction is lacking in the top soil, which may indicate a higher degree of leaching.

At site A there seems to be competition for light between seedlings on one hand and sprouts and *Chusquea* on the other. The competitive ability for light of the tree sprouts is considerably higher than that of tree seedlings as the sprouts grow from an established root system.

The density of seedlings, both in the primary forest and at site A, is low compared to tree sprouts and compared to site B, where roots, stumps, and rhizomes have been removed.

It seems, that at this location the ability of the montane rain forest to regenerate after a natural tree fall or a minor cutting is quite good, because of some important conditions:

- If roots, stumps and rhizomes are left at the regeneration site after a disturbance, sprouts become one of the most important groups in the regeneration path alongside with the bamboos.
- In most places *Chusquea* is an important factor because it creates a dense cover over the exposed soil soon after a clearing. The vegetation cover should be established immediately after an opening to prevent increased erosion as the slopes are extremely steep. Increased erosion could lead to land slides and with that, the removal of the top soil and the soil seed bank. The soil seed bank is important for the regeneration where roots, stumps and rhizomes have been removed. *Chusquea* is recruited from the adjacent primary forest by sprouting. Therefore the distance to primary forest and the size of the gap is important in determining the time, before a plant cover has been reestablished.

Floral scents in tropical pollination syndromes

Lars Tollsten and Jette Teilmann Knudsen

Dept. of Chemical Ecology, University of Göteborg, Box 33031, S-400 33 Göteborg, Sweden.

The classification by Fægri and van der Pijl (1979) and Baker and Hurd (1968) of flowers into different pollination syndromes based on floral morphology and color is well known. We intend to study floral odors to determine if flowers pollinated by one group of pollinators can be separated or recognized on the basis of floral odor compounds as well as based on morphology and color.

Our study of variation in floral odor will emphasize the following three questions:

1) Are there differences in floral odors between the pollination syndromes?
2) Is variation in floral odors mainly determined by taxonomic relations or by pollinator adaptations within closely related taxa?
3) Do generalistic flowers pollinated by a range of pollinators have less well defined floral odour compounds compared to specialized flowers?

Beetle, bat, and bird pollination syndromes will mainly be studied in the tropics (Ecuador), whereas bee, moth, butterfly, and fly syndromes will be studied mainly in temperate areas.

Samples of floral odors will be collected in the field by adsorption of scent on synthetic polymers and analyzed by GC-MS (Gas Chromathography-Mass Spectrometry). Three replicates will be taken from each plant species and the five largest components in each sample will be used for comparison of floral odor compounds. Samples will mainly be collected from plant species with known pollination biology but we will also make observations in the field.

Of special interest are families or genera where two or more different pollination syndromes occur (*e. g.*, Bignoniaceae, Gesneriaceae).

References

Baker, H. G. and Hurd, P. D. 1968. Intrafloral ecology. – Ann. Rev. Ent. 13: 385-414.

Fægri, K. and Pijl, L. van der 1979. The principles of pollination ecology (3. ed). – Pergamon Press, Oxford.

Pollination biology of the subtribe Bactridinae (Arecaceae)

Finn Borchsenius

Botanical Institute, University of Aarhus, Nordlandsvej 68, DK 8240 Risskov, Denmark

Pollination biology has been described for some 50 species af palms, belonging to 35 genera. In addition, the literature contains casual observations of flowering behaviour and insect visitors of 30 species belonging to an extra 28 genera. These studies show that palms are predominantly insect pollinated, whereas wind pollination, contrary to earlier beliefs, appears to be a rare condition. Other pollinators include bats and perhaps birds. The literature on palm pollination was reviewed by Henderson (1986).

Among insect pollinated palms, three common syndromes are found, each associated with a number of morphological and phenological characteristics. These three syndromes are: *Beetle pollination,* associated with phenomena such as protogyny, heating-up of buds before anthesis, nocturnal and short lived anthesis, and no nectar production. *Bee pollination,* associated with phenomena such as protandry, diurnal anthesis, staminate anthesis lasting one week or more, pistillate anthesis temporally separated and relatively short-lived, typically 2–3 days, and with nectar production. *Fly pollination,* associated with the same phenomena as bee pollination, but more common in understorey palms, and generally with a slower development of the inflorescences.

The subtribe Bactridinae is characterized by the presence of spines on most or at least some parts in all species. It is strictly neotropical, distributed from Mexico and the West Indies to Bolivia and northern Argentina. At present, six clearly demarcated genera are recognized within the group: *Acrocomia, Gastrococos, Aiphanes, Bactris, Desmoncus,* and *Astrocaryum* (Uhl and Dransfield 1987).

The species of *Acrocomia* are found in dryer lowland areas, avoiding ever-wet regions. Recently, studies of *Acrocomia acuelata* in Brazil have shown that this species has protogynous inflorescences, predominantly nocturnal anthesis, and is pollinated by weevils (Scariot 1987).

The monotypic genus *Gastrococos* is endemic to Cuba. There is no information about phenology or pollination in the genus.

Aiphanes is predominantly Andean, and most species are found in montane rain forest. The genus is characterized by praemorse pinnae, and protandrous inflorescences with a very slender peduncular bract. There are no studies on the pollination biology of this genus, but the phenomena associated with flowering correspond to the bee or fly pollination syndromes.

Bactris is by far the largest of the genera, with 239 recognized species. The members are predominantly understorey palms in lowland or premontane rain forest with a major concentration of species in Amazonian Brazil. Inflorescences are protogynous, with a relatively short peduncle and a thick, very broad, densely spinose peduncular bract. Beetle pollination has been demonstrated in several species (Henderson 1986).

The species of *Desmoncus* are climbers, and form an ecological parallel to the rattans of southeast Asia. Some species of *Desmoncus* have protogynous inflorescences strongly resembling those of *Bactris*, and we may expect that these display a similar pollination syndrome.

Astrocaryum is distributed in lowland rain forest, below 1,000 m altitude. Protogyny and beetle pollination have been demonstrated in two species.

The genera of Bactridinae differ in morphology, habit, and in terms of phenology and pollination mechanisms. *Aiphanes* appears to represent a distinct evolutionary line, characterized by praemorse pinnae and protandry, the latter probably as an adaptation to bee or fly pollination. This leads to the speculation that a different pollination mechanism was one of the major factors enabling *Aiphanes* to colonize the montane rain forests above 1,000 m altitude on a larger scale, as the only genus in Bactridinae. Indeed it seems that most of the dominant palm genera in these forests, *e.g.*, *Geonoma*, *Chamaedorea*, and *Prestoea* are all bee or fly pollinated, whereas genera known to be beetle pollinated with few exceptions are restricted to lowland areas.

References

Henderson, A. 1986. A review of pollination studies in the Palmae. – Bot. Rev. 52: 221–252.

Scariot, A. O. 1987. Biologia reproductiva de *Acrocomia aculeata* (Jacquin) Loddijes ex Martius (Palmae) do Distrito Federal. – Thesis. Univ. de Brasília, Brazil.

Uhl, N. W. and Dransfield, J. 1987. Genera Palmarum. – Allen Press, Lawrence, Kansas.

Pollination biology of the Carludovicoideae (Cyclanthaceae)

Roger Eriksson

Botaniska Institutionen, University of Göteborg, Carl Skottsbergs Gata 22, S-413 19 Göteborg, Sweden

The Cyclanthaceae, which have an exclusively neotropical distribution, consist of perennial herbs, shrubs, vines, and epiphytes. The leaves are usually bifid, seldom entire or palmately divided, and the inflorescence is a peduncled spadix subtended by spathes. The family forms a conspicuous part of the flora in humid vegetation types. The Cyclanthaceae comprise twelve genera in two subfamilies, *viz.* Cyclanthoideae (monotypic) and Carludovicoideae.

Cyclanthus bipartitus (Cyclanthoideae) has an inflorescence with pistillate and staminate flowers arranged in alternate cycles. The spathes produce a specialized tissue that serves as food source for the pollinators (in contrast to the members of Carludovicoideae). The spadix is protogynous and the pollinators, beetles of the genus *Cyclocephala* (Scarabaeidae), arrive when it is in the pistillate phase and leave at the end of the staminate phase, covered with pollen. The scarabs use the inflorescence as mating site.

The genera of Carludovicoideae (*Asplundia*, *Carludovica*, *Chorigyne*, *Dianthoveus*, *Dicranopygium*, *Evodianthus*, *Ludovia*, *Schultesiophytum*, *Sphaeradenia*, *Stelestylis*, *Thoracocarpus*) are characterized by an inflorescence with flowers in spirally arranged groups. The pistillate flowers have four usually distinct tepals (sometimes provided with an apical glandule), and four staminodes that are basally adnate to the tepals. The unilocular ovaries are protruding or sunken into the rachis. Styles are absent or four in number (often concrescent), and the stigmas are alternating with the tepals. Each pistillate flower is surrounded by four staminate ones, opposite the tepals. The staminate flowers, which usually are pedicelled, mostly have a perianth developed abaxially or all around the receptacle. The perianth lobes are usually provided with an external glandule producing a fat-containing substance. The stamens are more or less crowded on the receptacle, and in some genera the anthers are apically provided with a secretion globule. In many species the pollen is covered with pollenkitt. During anthesis the staminate flowers cover the pistillate ones, and only the anthers and staminodes, which protrude between the staminate flowers, are visible.

Field studies of *Asplundia*, *Dicranopygium*, and *Sphaeradenia* have revealed the same pattern of pollination biology, which probably is valid for most or all members of the Carludovicoideae.

During the first two days the spathes and staminodes successively unfold. Early in the morning (or during the night) of the third day, weevils (Curculionidae) arrive at the inflorescence, which has a higher temperature than the surrounding air. The attractant is a strong scent produced by the staminodes. At this stage the stigmas are receptive, but the anthers unopened, and hence the inflorescence is protogynous. The weevils enter between the staminate flowers, where the staminodes pass, and seem to bite off the staminodes. Often they enter two and two, probably for mating, and aggressively guard their position against other individuals. After that, they sit in the cavity above the pistillate flowers with their snout pointing out. Probably they are laying eggs, as one often finds larvae or adult individuals in the ovaries of mature infructescences. The perianth lobe glandules, which are exposed in the cavity, are possibly consumed by the beetles. This situation remains more or less unchanged all the day and evening. The fourth day, the anthers open, covering the weevils with pollen as they leave the inflorescence.

The spadix morphology and attracting scent of the Carludovicoideae restrict the effective pollinators to weevils, while these use the inflorescence as mating site and for laying eggs. This suggests a very close co-evolution between these cyclanths and their pollinators.

The evolution of dioecy in *Clavija* (Theophrastaceae)

Bertil Ståhl

Botaniska Institutionen, University of Göteborg, Carl Skottsbergs Gata 22, S-41319 Göteborg, Sweden

The genus *Clavija* comprises some 50 species of unbranched or sparsely branched shrubs and small trees. It is distributed mainly in the South American continent, inhabiting both deciduous forests and rain forests. Whereas a few species are hermaphrodite, the large majority are sexually polymorphic. The latter group includes species that are gynodioecious (female and hermaphrodite plants), polygamous (female, male, and hermaphrodite plants), androdioecious (male and hermaphrodite plants), and dioecious (male and female plants).

Schematically, it is suggested that the gynodioecious and polygamous conditions are intermediate stages of the evolutionary pathway from hermaphroditism to dioecy. These systems would then have arisen after establishment of male sterile mutants followed by a gradual reduction in ovule number in hermaphrodites. The androdioecious condition may have arisen directly from hermaphroditism by establishment of female-sterile mutants and/or after exclusion of female plants in polygamous populations.

The remaining, strictly hermaphrodite genera of the Theophrastaceae have an Antillean-centered distribution and occur mainly in dry to very dry areas. It is therefore argued that the evolution towards dioecy in *Clavija* could be seen as a part of its adaptive radiation in continental South America.

The constraints imposed by the moister and denser forest habitats may have included different pollinators and a higher cost for seed production. With the right pollen vectors present, separation of sexes may have been an advantageous way to allocate available resources to seeds capable of establishing new offsprings in a denser habitat.

Jukka Salo

Quarternary landscape processes in western Amazonas

Matti Räsänen

Department of Quarternary Geology, University of Turku, SF-20500 Turku, Finland

-

Jukka Salo

Department of Biology, University of Turku, SF-20500 Turku, Finland

The uppermost sedimentary layers which form the present dissected *terra firme* rain forest beds in the Amazon basin have in the western and central parts of the basin been described to form a relatively thin (10–40 m thick) flat-lying surficial geomorphological or structural formation. In the lowlands of western Amazonas, the uppermost layers of *terra firme* level appear to be formed by young (Pliocene-Holocene) sediments of Andean origin (Peru: Räsänen *et al*. 1987) or by older (Oligo-Miocene) sediments with an origin in the Guiana shield (Colombia: Hoon 1988). In both cases, the sedimentary environment has been predominantly fluvial.

In the Peruvian Amazonas, the active eastern Andean foreland dynamics have been accelerated by the shallow subduction of the marine Nazca plate. These foreland dynamics have affected the Plio-Pleistocene lowland relief evolution by 1) maintaining a system of temporarily shifting fluvial aggradational and degradational landscapes, 2) supporting further deformation of the aggraded plains, and 3) resulting in an altered hydrology with large-scale regional degradation, river capturing, river relocation and terrace formation.

The aggradational history of the non-flooded *terra firme* relief as well as the postdepositional development is of special interest for the biogeography of the Amazon basin. The biogeographic reasoning of Amazonian lowlands still in many cases relies on the "Belterra-lake" hypothesis, suggesting the central Amazonian plains to have been formed as a result of Calabrian lake sedimentation, and considers the present river drainage to have eroded the floodplain valleys into these sediments. The newly emerging view on fluvial aggradational history in western Amazonas suggests that the forest bed is inherently a mosaic structure. Due to young depositional age in the western Amazonas, many of the primary sedimentary facies of *terra firme* beds may have remained relatively well preserved and the floristic patterns of the forests may reflect these edaphic differences.

References

Hoon, C. 1988. Nota preliminar sobre la edad de los sedimentos Terciarios de la zona de Araracuara (Amazonas). – Bol. Geol. 29: 87.

Räsänen, M. E., Salo, J. S. and Kalliola, R. 1987. Fluvial perturbance in the western Amazon basin: regulation by long-term sub-Andean tectonics. – Science 238: 1398.

Dynamics of open and forested wetlands in Peruvian Amazonas

Risto Kalliola

Department of Biology, University of Turku, SF-20500 Turku, Finland

Most of the ecological studies on floodplain dynamics in the Amazon region have been carried out along the rivers Solimões and Negro in the vicinity of Manaus, in Brazil. These studies have clearly shown the dependence of all floodplain organisms on inundation cycles. The floodplain environments are both floristically and faunistically different from the surrounding unflooded *terra firme* areas. The ecological effects of the regularity of the aquatic and terrestrial phases in floodplain environments are often compared to the effects of seasonality in high latitutes. Consequently the floodplains in the central Amazonas are highly dynamic environments, but they are usually non-migrating as such.

In the upper Amazon region, in turn, most of the rain forests grow on a fluviodynamic mosaic, which is repeatedly modified by modern fluvial perturbance. In addition to channel migration, a number of other fluvial processes such as floodplain abandoning, river damming, and backswamp dynamics are present. The ecological consequences of these factors are often seen in the form of open and forested wetlands, which characterize especially the Pastaza, Marañon, and Ucayali basin areas.

The wide presence of flooded areas, the large variation in flood amplitudes, and the ever changing nature of vegetation, all suggest several different levels in flood adaptation of species. A great number of descriptive and experimental work, as well as floristical inventories are needed in order to understand the complexity of the different biological interactions occurring here.

Preliminary results from studies on the vegetation of the floodplain forest of Peruvian Amazonas.

Maarit Puhakka

Department of Biology, University of Turku, SF-20500 Turku, Finland

Perturbance processes promoted by fluvial activity have been the major factors causing large scale forest destruction and the subsequent primary succession in the western lowland Amazonia. Large areas of the region show signs of recent fluvial activity and most of the forested areas have gone trough riparian primary succession at least once in their lifetime. Studies on these phenomena are essential for understanding the natural rainforest dynamics in the region.

We have conducted a field program for studying sequential succession of forest in the Peruvian Amazonas since 1987. The work is based on field work made in several different regions in order to cover all possible variations in the successional features of the study area. An extensive set of information has been collected concerning species composition, forest structure, fluvial landforms, and soil chemistry, using line transect analyses. The study includes descriptions of successional zones (structure) with their floras, and the variation of the successional patterns at different rivers and regions.

The data has been collected at eight rivers between the western latitudes 69°–75°. We have five different meander loops in the southern Peru in Madre de Dios basin at the rivers Pinquen, Manu, Madre de Dios and Malinowski, two in the Ucayali basin at the rivers Ucayali and Tapiche, two near the margin of the Pastaza-Marañon at the river Tigre, and finally two at the river Nanay, tributary of the river Amazonas. The rivers are grouped in active white and more stable mixed water rivers (the rivers Nanay, Tapiche, and Tigre).

The major part of the information was collected using one to three line transect analyses at each site (the number of the transects totals 17). All the data were collected during the dry and low water seasons (July–October) of 1987–1988. Each transect was set up against the age gradient of the successional forest. After this, the transect distances were measured from the current tip of the point bar, and elevations (topography) were measured, using the current water level as a base line.

Information on trees was collected within a four meters wide corridor. The position of each tree was determined by using distance from the river and the elevation. The tree individuals were then numbered, their heights were estimated and DBH was measured. In addition, various structural information was collected from each individual.

All the trees studied were identified in the field using temporary names. One to two voucher specimen were collected under each of these names, giving a total of 645 voucher specimens. Most of the material was later identified at the Botanical Institute of Aarhus University (AAU).

The data consists of 3362 individual trees, of which 72 (2.1%) remain unidentified. There are 52 families, 125 genera, and 379 species in the entire material studied; 14 families include 10 species or more, and together these comprise 69.7% of the species identified.

The families with most species are: Rubiaceae, Euphorbiaceae, Solanaceae, and Annonaceae. Families such as Acanthaceae and Elaeocarpaceae occured only along white water rivers, while the families Myrsinaceae, and Sapotaceae only occurred in mixed water environments.

The 22 most common species account for 50% of the individuals studied, most of them belonging to early stages of the succession. Of all the species identified 7.9% occurred along both white and mixed water rivers.

Ecology of a rust fungus along Río Manu in Peru

Yrjö Mäkinen

Department of Biology, University of Turku, SF-20500 Turku, Finland

The ecology and occurrence of the seedlings of *Tessaria integrifolia* along Río Manu in the province of Madre de Dios (Peru) are described, together with ecology and occurrence of a parasitic rust fungus, *Uromyces megalospermus*. The rust occurs only on seedlings and as a systemic parasite it causes severe deformations. In infection experiments healthy leaves were more easily infected than the infected leaves; this points to chemical defense of the host. Also various ecological conditions may contribute to the resistence of the mature leaves. The evolutionary significance of the host-parasite relationship is discussed.

Occurrence of vesicular-arbuscular mycorrhiza (VAM) in the seasonally flooded forest of the Mapire River (Anzoategui, Venezuela)

*Tania M. de la Rosa *, G. Cuenca and R. Herrera*

Centro de Ecología, Instituto Venezuelano de Investigationes Científicas, Apartado 1827, Caracas, Venezuela

**Present address: Department of Biology, University of Turku, SF-20500 Turku, Finland*

Tropical wetlands have recently become of great interest for ecological studies, but still very little is known about the biotic and abiotic processes which are due to fluctuations in water level (Herrera *et al*. 1986). This project is part of an integrated ecological study carried out in the seasonally flooded forest of the Mapire River, and describes the occurrence of vesicular-arbuscular mycorrhizae (VAM). The VAM are symbiotic, mutualistic associations established between the root systems of most plants and certain fungi. They increase the uptake of soil nutrients, especially in oligotrophic tropical soils.

Until about 15 years ago, the literature (Gerdemann 1968; Khan 1974) considered those plants of continuous or periodically inundated habitats as non-mycorrhizal. However, VAM has recently been found in a number of submerged, freshwater aquatic plants (Søndergaard and Lægaard 1977; Clayton and Bagyaraj 1984; Tanner and Clayton 1985), and in temperate swamps (Keeley 1980). In inundated conditions, the flooding of the soil affects the growth of plants and their root system (Kozlowski 1984), but the presence of VAM could improve the uptake of nutrients.

A general characterization of the seasonally flooded forest area was carried out. Two zones were distinguished according to the degree of inundation (water level and time): high inundated zone and low inundated zone. The main objectives for the studies were:

1) the quantification of the percentage of VAM infection.
2) determination of the mycorrhizal status of some of the tree species.
3) the identification of the fungal species involved in the symbiosis.

The analysis of the soil showed that its acidity (pH 4,3) and its low content of phosphorus (< 2 ppm) would favor the presence of the VAM. Most of the woody tree species considered (93%) show varying degrees of mycorrhizal infection. Several fungi belonging to four different genera have been identified.

This study shows that the presence and amount of VAM in the seasonally inundated forest of the Mapire River is similar to that observed in other natural terrestrial or aquatic ecosystems. It also confirms their capacity to develop under alternating dry and inundated conditions.

References

Clayton, J. S. and Bagyaraj D. J. 1984. Vesicular arbuscular mycorrhizas in submerged aquatic plants of New Zealand. – Aquatic Botany 19: 251–262

Gerdemann, J. W. 1968. Vesicular-arbuscular mycorrhiza and plant growth.– Ann. Rev. Phytopat. 6: 397–418

Herrera, R., Añez, M. A., Barrios, E., Flores, S., Isquirdo, L., de la Rosa, T., Rosales, J., Valles, J.,L., and Vegas, T. 1986. Nutrient cycling (spiralling) and forest dynamics in a seasonally flooded forest in Middle Orinoco. – International workshop on tropical rain forest regeneration and management. Guri, Venezuela

Keeley, J. E. 1980. Endomycorrhizae influence on growth of Blackgum seedlings in flooded soils. – Amer. J. Bot. 67(1): 6–9

Khan, A. G. 1974. The occurrence of mycorrhizas in halophytes, hydrophytes and xerophytes, and of *Endogone* spores in adjacent soils. – J. Gen. Microbiol. 81: 7–14

Kozlowski, T. T. 1984. Plant responses to flooding of soil. – BioScience 34(3): 162–167

Søndergaard, M. and Laegaard, S. 1977. Vesicular-arbuscular mycorrhiza in some aquatic vascular plants. – Nature 268: 232–233

Tanner, C. C. and J. S. Clayton, J. S. 1985. Vesicular arbuscular mycorrhiza studies with a submerged aquatic plant. – Trans. Br. Mycol. Soc. 85 (4): 683–688

Confocal Scanning Laser Microscopy, a new technique used for embryological studies in orchids

Margit Frederiksson

Botaniska Institutionen, University of Göteborg, Carl Skottsbergs Gata 22, S-413 19 Göteborg, Sweden

Confocal scanning offers a possibility for three-dimensional microscopy, since a recording of the entire structure of an ovule can be made by scanning a number of confocal images, and refocusing the microscope between successive images. The result is a stack of images representing the three-dimensional structure of the ovule. Optical serial sectioning can be done quickly.

For confocal microscopy to be successfully applied, the specimen must be reasonably transparent to allow light to penetrate to regions below the surface. This is achieved by using an ordinary clearing technique.

The Aarhus University Ecuador project 1987–1989

Finn Borchsenius
Botanical Institute, Aarhus University, Nordlandsvej 68, DK 8240 Risskov, Denmark

Flora of Ecuador

Taxonomic studies for the Flora of Ecuador in the period have been concentrated mainly on four major groups: Pteridophytes (especially Lycopodiaceae), Poaceae, Palmae, and Leguminosae. Lycopodiaceae was published by B. Øllgaard in 1988. More than half the genera of Ecuadorean palms have been worked out, and little work remains do be done on most of the rest (H. Balslev and coworkers). In 1988, Ivan Nielsen became coordinator of the Leguminosae for the Flora of Ecuador. Studies of *Acacia* (E. B. Madsen) and *Brownea* (B. Bang-Klitgaard) have been subjects for M. Sc. theses.

Vegetation studies

The study of forest composition and structure in Añangu, Amazonian Ecuador was concluded in 1987, and the results have been published (Balslev *et al.* 1987).

A study of the natural vegetation on Isla Puná funded by DANIDA was undertaken by J. E. Madsen in 1987 under the direction H. Balslev and in collaboration with Banco Central, Ecuador. The project was concluded and reported on in 1988 (Balslev *et al.*).

A new three year project concerning the diversity and dynamics of lowland rain forest in Reserva Faunistica Cuyabeno was launched in 1987 by H. Balslev *et al.*, in collaboration with P. Universidad Católica del Ecuador (PUCE). Guillermo Paz y Miño, Renato Valencia (PUCE) are employed in the project, and several students have participated (A. D. Poulsen, I. Nielsen , H. Christensen, K. Bloch).

A DANIDA funded project concerning diversity and regeneration of montane forest in Parque Nacional de Podocarpus in southern Ecuador was initiated in 1988 under the direction of B. Øllgaard and in collaboration with PREDESUR and Universidad Nacional de Loja. Field work has been undertaken by J. E. Madsen and several students have participated in the investigations (L. Christensen, Bj. Eriksen, A. B. Pedersen, J. P. Feil).

A study of species composition and present extension of montane forest above 2,400 m in Ecuador continues under the direction of P. M. Jørgensen and C. Ulloa.

Field work

In the period 1987–1989 members of the Aarhus group have undertaken several collecting expeditions to Ecuador, with different scopes:

Systematic expeditions related to ongoing revisional work: *Fungi* (Læssøe 1987); *Polygonaceae* (Brandbyge 1987); *Poaceae* (Lægaard and Renvoize 1988); *Palmae* (Barfod *et al.* 1987, Balslev *et al.* 1987, Skov and Borchsenius 1987, Blicher Mathiesen 1987, Bergmann and Pedersen 1988, Borchsenius 1989); *Leguminosae* (Bang Klitgaard 1987, Bjerrum Madsen 1988); *Melastomataceae* (Renner 1988). *Gesneriaceae* (Kvist 1986–1987).

General collections related to vegetation studies: *Isla Puná*, Guayas province (Madsen 1987); *Cuyabeno*, Sucumbios province (Balslev *et al.* 1987, 1988, 1989); *Parque Nacional de Podocarpus*, Loja province (Madsen *et al.* 1988–89, Øllgaard 1988, 1989); *Montane forest*, especially at Pasochoa and Lloa in the Pichincha province (Jørgensen *et al.* 1986–1989); Ethnobotanical collections: *Saraguro Indians*, Loja province (Ellemann 1988–1989).

The overall number of AAU collections from Ecuador in the period adds up to 30,000, which means that the total number of AAU collections from Ecuador is now more than 90,000.

In addition to field work in Ecuador, AAU staff members have participated in a number of expeditions to other neotropical countries, concentrating on specific groups. *Palmae*: Moraes and Balslev 1987, Bolivia; Barfod 1987, 1988, Peru, Colombia, Panama; Blicher-Mathiesen 1987, Peru; Borchsenius 1989, Colombia. *Poaceae*: Lægaard 1988, Peru. *Gesneriaceae*: Kvist 1987, Colombia, Guyana. *Melastomataceae*. Renner 1989, Brazil.

International meetings

In August 1988 Botanical Institute arranged an international symposium on "Tropical forests: Botanical dynamics, Speciation and Diversity" to celebrate the 25th anniversary of the institute. The abstracts for the presentations were published as an issue of AAU Reports (Barfod and Skov 1988). The proceedings of the symposium were published by Academic Press (L. B. Holm-Nielsen *et al.* 1989).

Collaboration with P. Universidad Católica (PUCE) in Quito

The close relationship between Aarhus and PUCE is still evolving. Three members of the Aarhus group have been residents in Quito in the period: Peter Møller Jørgensen (1986–1989), Henrik Borgtoft Pedersen and Birgitte Bergmann (1989–). In 1990 a new long term project concerning scientific collaboration between PUCE and Aarhus University was intitated. The project is financed by DANIDA and B. Øllgaard has been appointed guest professor at PUCE for the period 1990-1992.

Three Ecuadorean students have had fellowships in Aarhus in 1987-1989: Guillermo Paz y Miño (1987), Carmen Ulloa (1988–1989), and Katya Romoleroux (1989–1990).

Research program

1. TAXONOMY

a. Flora of Ecuador

Alismatidae (L. B. Holm Nielsen and R. R. Haynes, published 1986).
Anacardiaceae (A. Barfod, published 1987).
Passifloraceae (L. B. Holm Nielsen, P. M. Jørgensen and J. E. Lawesson, published 1988).
Lycopodiaceae (B. Øllgaard, published 1988).
Cactaceae (J. E. Madsen, published 1989).
Valerianaceae (B. Eriksen and B. B. Larsen, published 1989).
Polygonaceae (J. Brandbyge, published 1989).
Poaceae (S. Lægaard coordinator, I. Grignon, K. Gludsted, U. Hjort, S. Renvoize, E. Judziewicz, L. Clark).
Palmae (H. Balslev, A. Barfod, F. Skov, B. Bergmann, H. Borgtoft Pedersen, U. Blicher-Mathiesen, F. Borchsenius).
Leguminosae (I. Nielsen coordinator, B. Bang Klitgaard, E. B. Madsen).
Monimiaceae (S. S. Renner).
Rosaceae (K. Romoleroux).
Pteridaceae (A. L. Arbelaez).

b. Revisions

Triplaris (Polygonaceae, J. Brandbyge, published 1986).
Aciachne (Poaceae, S. Lægaard,published 1987).
Hyospathe (Palmae, F. Skov, published 1989).
Rhyncanthera (Melastomataceae, S. S. Renner, published 1989).
Macairea, Bellucia, Loreya (Melastomataceae, S. S. Renner, published 1990).

Pterogaster, Pterolepis, Schwarzea (Melastomataceae, S. S. Renner).
Phytelephantoideae (Palmae, A. Barfod, submitted).
Muehlenbeckia (Polygonaceae, J. Brandbyge, submitted).
Aiphanes (Palmae, F. Borchsenius).

2. FLORISTIC STUDIES

a. Inventories

Oriente checklist (S. S. Renner, H. Balslev, L. B. Holm-Nielsen).
Upper Andean forest (P. M. Jørgensen, C. Ulloa).

b. Forest diversity and dynamics

Amazonian rain forest (H. Balslev *et al.*). Collaborator PUCE.
Montane forest (J. E. Madsen, B. Øllgaard, P. M. Jørgensen). Collaborators PUCE, DANIDA, PREDESUR, UNL.

3. ETHNOBOTANY

Uses of Ecuadorean palms (H. Balslev, A. Barfod, H. Borgtoft Pedersen).
Ethnobotany of the Saraguro Indians (L. Ellemann).
Firewood (B. Øllgaard, J. E. Madsen).

Participants in the Aarhus University Ecuador Project

1. PERMANENT STAFF

Henrik Balslev, taxonomy of Juncaceae and Palmae, vegetation studies of western Amazonian rain forest, ethnobotany.

Lauritz B. Holm-Nielsen, taxonomy of Alismatidae and Passifloraceae, phytogeography, ethnobotany (at present on leave of absence serving as Rector of the Danish Research Academy).

Simon Lægaard, taxonomy of Poaceae, páramo vegetation.

Ivan Nielsen, taxonomy of Mimosaceae.

Benjamin Øllgaard, taxonomy of Pteridophytes, especially Lycopodiaceae, vegetation studies of montane forest.

2. ASSOCIATED STAFF

Anders Barfod, taxonomy of Anacardiaceae and Palmae, ethnobotany.

Susanne S. Renner, taxonomy of Melastomataceae and Monimiaceae, vegetation studies, pollination biology.

3. STUDENTS

a. Ph. D. Students

1986–88 Anders Barfod. The natural history of Phytelephantoideae (Palmae). Funded by The Danish Natural Science Reseach Council.

1987–89 John Brandbyge. Systematic studies of neotropical Polygonaceae. Funded by Aarhus University.

1987–89 Flemming Skov. Use of computers in taxonomy, *Geonoma* (Palmae) for Flora of Ecuador. Funded by Aarhus University.

1989–91 Finn Borchsenius. Taxonomy and pollination biology of *Aiphanes* (Palmae). Funded by Aarhus University.

1989–91 Peter Møller Jørgensen. Species composition and present extension of Andean forest above 2,400 m altitude in Ecuador. Funded by DANIDA.

1989–91 Jens Elgaard Madsen. Species composition, dynamics, and regeneration of montane forest in Parque Nacional de Podocarpus in Ecuador. Funded by DANIDA.

b. Master degree students

Completed in 1987

Bente Eriksen. *Valeriana* in an Ecuador (supervisor L. B. Holm-Nielsen)

Jørgen Korning. Composition of an Ecuadorean lowland rain forest (supervisor B. Øllgaard).

Karsten Thomsen. Compostion of Ecuadorean lowland rain forest (supervisor B. Øllgaard).

Jens Madsen. Cactaceae in Ecuador (supervisor L. B. Holm-Nielsen).

Completed in 1988

Isabelle Grignon. The genus *Sporobolus* in Ecuador (supervisor S. Lægaard).

Kirsten Gludsted. *Cortaderia* (Poaceae) in Ecuador (supervisor S. Lægaard)

Finn Borchsenius. The genus *Aiphanes* Willd. in Ecuador (supervisor H. Balslev).

Ulla Blicher-Mathiesen. Attaleinae (Palmae) in Ecuador (supervisor H. Balslev).

Henrik Borgtoft Pedersen. Biology, utilization and management of Ecuadorean palms (supervisor H. Balslev).

Completed in 1989

Birgitte Bergmann. A taxonomic revision of *Chamaedorea* in Ecuador (supervisor H. Balslev).

Monica Moraes. A revision of *Allagoptera* (Palmae) (supervisor H. Balslev).

Lars Christensen. Forest regeneration in Parque Nacional de Podocarpus (supervisor B. Øllgaard).

To be completed in 1990

Klaus Bloch. A tree inventory in a one-hectare plot in tropical rain forest in Ecuador (supervisor H. Balslev).

Bente Bang Klitgaard. *Brownea* and *Browneopsis* in Ecuador (supervisor I. Nielsen).

Ingvar Nielsen. Epiphyte vegetation in a one-hectare plot in tropical rain forest in Ecuador (supervisor H. Balslev).

Eva Bjerrum Madsen. *Acacia* in Ecuador (supervisor I. Nielsen).

Axel Dahlberg Poulsen. A quantitative and qualitative inventory of the herbaceous ground vegetation in a one-hectare plot in tropical rain forest in Ecuador (supervisor H. Balslev).

Lis Elleman. Ethnobotany of the Saraguro indians in Ecuador (supervisor B. Øllgaard).

Henning Christensen. Vegetation studies in tropical rainforest in Ecuador (supervisor H. Balslev).

Bjarke Eriksen. A quantitative and qualitative inventory of the herbaceous ground vegetation in a one-hectare plot in montane forest in Parque Nacional de Podocarpus, Ecuador (supervisor B. Øllgaard).

Anders Bøgh Pedersen. Epiphyte vegetation in a one-hectare plot in montane forest in Parque Nacional de Podocarpus, Ecuador (supervisor B. Øllgaard).

Ulla Hjorth. The genus *Poa* in Ecuador (supervisor S. Lægaard).

Jan Peter Feil. Reproductive ecology of Ecuadorean Monimiaceae (supervisor S. S. Renner).

Dorte Ellinor Christensen. Pollination and fruit set in *Stelis argentata* Lindl. (Orchidaceae) (supervisor S. S. Renner).

Birgitte Mogensen. Maya ethnobotany in Yucatan (supervisor H. Balslev)

Ole Stauning. Maya ethnobotany in Yucatan (supervisor H. Balslev).

c. Fellowships in Aarhus

1987 Roger Eriksson, University of Göteborg, Sweden. Cyclanthaceae. Funded by Aarhus University, Nordic Stipendium.

1988 Guillermo Paz y Miño, PUCE, Ecuador. Flora of Cotopaxi, Ecuador. Funded by Botanical Institute.

1988–89 Monica Moraes, Herbario Nacional, La Paz, Bolivia. Bolivian Palms. Funded by Danish Research Academy.

1988–89 Carmen Ulloa, PUCE, Ecuador. Montane forest in Ecuador. Funded by DANIDA.

1988–89 Bertil Ståhl, University of Göteborg. Theophrastaceae. Funded by Nordic council, research grants.

1989 Steve Churchill, The New York Botanical Garden. Mosses of Ecuador and Colombia. Funded by The Danish Research Academy.

1989 Alba Luz Arbalaez, Universidad de Antiochia, Colombia. *Pteris* in Ecuador. Funded Botanical Institute, Aarhus University.

1989–90 Risto Heikkinen, University of Turku, Finland. Tropical ecology. Funded by Aarhus University, Nordic Stipendium.

1989–90 Katya Romoleroux, PUCE, Ecuador. Ecuadorean Rosaceae. Funded by DANIDA.

Publications on neotropical botany 1987–1989

Balslev, H. 1987: Palmas nativas de la Amazonia Ecuatoriana. – Revista Colibri 3: 64–73.

– 1988a: Distribution patterns of Ecuadorean plant species. – Taxon 37: 567–577.

– 1988b: Two new rushes (*Juncus*, Juncaceae) from Chiapas. – Ann. Missouri Bot. Gard. 75: 379–382.

– 1988c: Scientific collaboration between Universidad Católica del Ecuador and Aarhus University. – In: Dolberg, F. and Østergaard, C. L. (eds.), Directions and new working procedures in rural development research: 37–47. Institute of political science, Aarhus University.

– 1988d: Ecuador - et biologisk smørhul. – Udvikling 6: 31–33.

– and Barfod, A. 1987: Ecudorean Palms – an overview. – Opera Bot. 92: 17–35.

– and Henderson, A. 1987a: A new *Amandra* (Arecaceae) from Ecuador. – Syst. Bot. 12: 501–504.

– and Henderson, A. 1987b: *Elaeis oleifera* (Palmae) encontrada en el Ecuador. – Publ. Mus. Cienc. Nat. Ecuador 5: 45–49.

– and Henderson, A. 1987c: The identity of *Ynesa colenda* O. F. Cook. – Brittonia 39: 1–6.

– and Henderson, A. 1987d: *Prestoea palmito*. – Principes 31: 11.

– and Moraes, M. 1989: Sinopsis de las palmeras de Bolivia. – AAU rep. 20: 1–107.

– and Renner, S. S. 1989: The diversity of Ecuadorean forests east of the Andes. – In: Holm-Nielsen, L. B., Nielsen, I. and Balslev, H. (eds.), Tropical forests: Botanical dynamics, Speciation and Diversity: 287–295. Academic Press, London.

– , Luteyn, J. L., Øllgaard, B., and Holm-Nielsen, L. B. 1987: Composition and structure of adjacent unflooded and floodplain forests in Amazonian Ecuador. – Opera Bot. 92: 37–57.

– , Madsen, J., and Mix, R. 1988: Las plantas y el hombre en la Isla Puná, Ecuador (67 pp.). – Universidad Laica Vicente Rocafuerte and Museo Antropologico, Guayaquil.

Barfod, A. 1988a: The evolutionary and phylogenetic significance of the inflorescence morphology of some South American Anacardiaceae. – Nord. J. Bot. 8: 3–11.

– 1988b: Pollen morphology of *Ammandra*, *Palandra* and *Phytelephas* (Arecaceae). – Grana 27: 239–242.

– 1989: The rise and fall of vegetable ivory. – Principes 33: 181–190.

– and Balslev, H. 1988: The use of Palms by the Cayapas and Coaiqueres on the coastal plain of Ecuador. – Principes 32: 29–42.

– , Henderson, A., and Balslev, H. 1987: A note on the pollination of *Phytelephas microcarpa* (Palmae). – Biotropica 19: 191–192.

– and Skov, F. (eds.) 1988: Tropical forests: botanical dynamics, speciation, and diversity. Abstracts from the AAU 25th anniversary symposium. – AAU reports 18: 1–46.

Borchsenius, F. and Balslev, H. 1989: Three new species of *Aiphanes* (Palmae) with notes on the genus in Ecuador. – Nord. J. Bot. 9: 383–393.

– , Bernal, R., and Ruiz, M. 1989: *Aiphanes tricuspidata* (Palmae), a new species from Colombia and Ecuador. – Brittonia 41: 156–159.

Brandbyge, J. 1989a: Polygonaceae. – In: Harling G. and Andersson, L., Flora of Ecuador 38: 1–62.

– 1989b: Notes on the genus *Ruprechtia* (Polygonaceae). – Nord. J. Bot. 9: 57–61.

– 1989c: Two new species of the genus *Coccoloba* (Polygonaceae). – Nord. J. Bot. 9: 205–208.

– and Holm-Nielsen, L. B. 1987: Reforestacion de los Andes Ecuatorianos con especies nativas. (Spanish translation of "Reforestation of the high Andes with local species", Rep. Bot. Inst. AU 13: 1–114). – C.E.S.A. Quito, Ecuador.

– and Rechinger, K. H. 1989: A new *Rumex* from Ecuador. – Nord. J. Bot. 9: 203–204.

Eriksen, B. 1989a: Valerianaceae. – In: Harling, G. and Andersson, L. (eds.), Flora of Ecuador 34: 1–60.

– 1989b: Notes on the generic and infrageneric delimitation in the Valerianaceae. – Nord. J. Bot. 9: 179–187.

Figueroa, E., Oviedo, G., Vela, C., Sierra, R., Balslev, H., Torres, J., Carrasco, A., and de Vries, T. 1987: Evaluación del impacto ambiental del sismo en la Amazonía (230 pp., with 7 figs. and 9 maps). – Fundación Natura, Quito.

Garcés, G. G. and Skov, F. 1989: *Geonoma linearis* – a rheophytic palm from Colombia and Ecuador. – Principes 33: 108–112.

Grignon, I. and Lægaard, S. 1988: *Muhlenbergia palmirensis*, a new species from Ecuador (Poaceae). – Nord. J. Bot. 8: 47–49.

Holm-Nielsen, L. B. and Lawesson, J. E. 1987: New species of *Passiflora* subg. *Passiflora* from Ecuador. – Ann. Missouri Bot. Gard. 74: 497–504.

– , Jørgensen, P. M. and Lawesson, J. E. 1988: Passifloraceae. – In: Harling, G., and Sparre B., Flora of Ecuador 31: 1–129.

– , Nielsen, I., and Balslev, H. (eds.) 1989: Tropical forests: Botanical dynamics, Speciation and Diversity. – Academic Press, London.

Jaramillo, J. L. and Jørgensen, P. M. 1988: Estudios botanicos sobre la taxonomía del bosque montano. – Mem. I Cong. Cienc. Ecuador II: 144–148.

Jørgensen, P. M. and Jaramillo, J. L. 1988: La Ceja Andina – Una vegetación que desaparece. – Mem. I Cong. Cienc. Ecuador II: 119–123.

– and Ulloa U. (eds.) 1989: Estudios botanicos en la "Reserva ENDESA", Pichincha – Ecuador. – AAU rep. 22: 1–138.

– and Valencia, R. 1988: Composición de un bosque Andino, Pasochoa – Ecuador. – Publ. Mus. Cienc. Nat. Ecuador 6: 21–38.

– , Holm-Nielsen, L. B., and Lawesson, J. E. 1987: New species of *Passiflora* subgenus *Plectostemma* and subg. *Tacsonia* (Passifloraceae). – Nord. J. Bot. 7: 127–133.

Kvist, L. P. 1987: Hunting *Kohleria* in Colombia. – The Gloxinian 37(4): 17–21.

– 1988a: Regnskovens Naturfolk. – In: Bitch, I. R. *et al.* (eds.). Alle Tiders Regnskov: 28–36. Mellemfolkeligt Samvirke, Copenhagen.

– 1988b: Rejseindtryk fra Guyana. – Regnskov 5(4): 15–17.

– 1988c: Manu Nationalpark. – In: Bitch, I. R. *et al.* (eds.). Alle tiders Regnskov: 86–88. Mellemfolkeligt Samvirke, Copenhagen.

– 1989: Popular names and medicinal uses of *Columnea* (Gesneriaceae). – The Gloxinian 39(2): 21–25.

– and Holm-Nielsen, L. B. 1987: Ethnobotanical aspects of Ecuador. – Opera Bot. 92: 83–107.

– and Skog, L. E. 1988a: *Columnea incredibilis* and *Cremosperma filicifolium*; two new and remarkable Gesneriaceae from Western Colombia. – Nord. J. Bot. 8: 253–257.

– and Skog, L. E. 1988b: The genus *Cremosperma* (Gesneriaceae) in Ecuador. – Nord. J. Bot. 8: 259–269.

– and Skog, L. E. 1988c: Revision of *Reldia* (Gesneriaceae). – Nord. J. Bot. 8: 601–611.

Lægaard, S. 1987: The genus *Aciachne* (Poaceae). – Nord. J. Bot. 7: 667–672.

Lawesson, J. E. 1987: Problems of plant conservation in the Galapagos Islands. – Noticias de Galapagos 44: 12–13.

– 1988a: The stand level dieback and regeneration of forests in the Galapagos Islands. – Vegetatio 77: 87–93.

– 1988b: Contributions to the flora of the Galapagos Islands, Ecuador. – Phytologia 65: 228–230.

– and Adsersen, H. 1987: Notes on the endemic genus *Darwiniothamnus* (Asteraceae - Astereae) from the Galapagos Islands. – Opera Bot. 92: 7–15.

– and Estupinan, A. 1987: The reforestation project on San Cristobal Island. – Noticias de Galapagos 45: 23.

– and Norman, E. 1987: Buddlejaceae, a new native family for Galapagos. – Noticias de Galapagos 45: 28.

– , Adsersen, H., and Bently, P. 1987: An updated and annotated checklist of the vascular plants of the Galapagos Island. – Rep. Bot. Inst. AAU 16: 1–74.

Lescure, J.-P., Balslev, H., and Alarcón, R. 1987: Plantas utiles de la Amazonia Ecuatoriana (407 pp.). – ORSTOM, PUCE, and PRONAREG, Quito.

Macdonald, I., Ortiz, L., Lawesson, J. E., and Nowak, J. B. 1988: The invasion of the highlands in Galapagos by the red quinine tree *Cinchona succirubra*. – Environmental Conservation 15: 215–220.
Madsen, J. E. 1987: Kaktus i junglen. – Regnskov 2: 16–17.
– 1989: Cactaceae. – In: Harling, G. and Andersson, L. (eds.), Flora of Ecuador 35: 1–79.
Norman, E. and Lawesson, J. E. 1987: *Buddleja americana* L. (Buddlejaceae). – Madroño 34: 90–91.
Nowak, J. B. and Lawesson, J. E. 1988: Revegetation of the burnt area in Isabella. – Noticias de Galapagos 46: 18–20.
Olesen, J. M. 1987: Bier og blomster. – Regnskov 2: 7–10.
– 1988: Nest structure of an *Euglossa* sp. nov. in a fruit of *Theobroma subincanum*, Ecuadorean Amazonas. – Acta Amaz. 18: 327–330.
– 1989: Nest structure of the Amazonian *Bombus transversalis* (Olivier). – J. Trop. Ecol. 5: 243–246.
Renner, S. S. 1987: Seed dispersal. – Progress in Botany 49: 413–432.
– 1989a: Systematic studies in the Melastomataceae: *Bellucia*, *Loreya*, and *Macairea*. – Mem. York Bot. Gard. 50: 1–112.
– 1989b: Floral biological observations on *Heliamphora tatei* (Sarraceniaceae) and other plants from Cerro de la Neblina in Venezuela. – Pl. Syst. Evol. 163: 21–29.
– 1989c: A survey of reproductive biology in neotropical Melastomataceae and Memecylaceae. – Ann. Missouri Bot. Gard. 76: 496–518.
– 1989d: Reproduction and evolution in some genera of neotropical Melastomataceae. – In: Prance, G. T., and Gottsberger, G. (eds.), The biology of tropical woody angiosperms. Mem. New York Bot. Gard. 55: 143–152.
– 1989e: The current classification of Melastomataceae. – Flora Malesiana Symposium, abstracts: 52.
– 1990: A revision of *Rhyncanthera* (Melastomataceae). – Nord. J. Bot. 9: 601–630.
Skov, F. 1989: HyperTaxonomy – a new computer tool for revisional work. – Taxon 38: 582–590.
– and Balslev, H. 1989: A revision of *Hyospathe* (Arecaceae). – Nord. J. Bot. 9: 189–202.
Øllgaard, B. 1987a: A revised classification of the Lycopodiaceae. – Opera Bot. 92: 153–178.
– 1987b: Bregner i Trætoppene. – Regnskov 4: 13–15.

– 1987c: Alexander von Humboldt og den botaniske udforskning af Amerika. – Naturens Verden 1987(1): 23–4.
– 1988: Lycopodiaceae. – In: Harling G. and Andersson, L., Flora of Ecuador 33: 1–155.
– 1988: Index of the Lycopodiaceae. – Biol. Skr. Dan. Vid. Selsk. 34: 1–135.
– and Windisch, P. G. 1987: Sinopse das Licopodiaceas do Brasil. – Bradea 5: 1–43.

Aknowledgements

Several funds, organisations, and private companies have supported our work financially: DANIDA; Danish Natural Science Research Council; Aarhus University Research Foundation; Danish Research Academy; Danish Research Council for Developmental Studies; Carlsberg Fund; Charles Lindberg Fund; Julie von Müllens Fund; Suzanne Liebermann Erichson Fund of the Smithsonian Institution; CONUEP, Ecuador; Latinreco S.A., Ecuador; Cononco Ltd., Ecuador. We are most greatful for this support.

The Göteborg University Ecuador project 1987–1989

Bente Eriksen

Botaniska Institutionen, University of Göteborg, Carl Skottsbergs Gata 22, S-413 19 Göteborg, Sweden

Flora of Ecuador

Most research projects at the Department of Systematic Botany in Göteborg cover more than the regional flora work of Ecuador. The Scitaminae-group (L. Andersson and M. Hagberg) has completed the Ecuadorean part of the work but exsists until the "Flora Neotropica" is ready for publishing. The Scrophulariaceae volume of the Flora has since long been published, and projects in the Scrophulariaceae-group (U. Molau and F. Astholm) are of revisional or more experimental character. The macrolichen-group (L. Arvidsson, M. Lindström, and M. Lindqvist) proceeds on the genera *Erioderma*, *Sticta*, and *Leptogium*. A new research group was established in 1989. This group, coordinated by L. Andersson, is going to revise the Neotropical Rubiaceae (see L. Andersson, this publ.). Besides the established project groups, a number of separate revisions are carried out on various plant families and genera (see "Staff and students ...").

Status of the Flora of Ecuador

Flora of Ecuador is continuously being published, for the time being under the editorship of prof. emeritus Gunnar Harling and prof. Lennart Andersson. M. Sc. Bertil Ståhl is assistant editor. Publication of the Flora of Ecuador is sponsored by the Swedish Natural Science Research Council.

For a list of volumes published in Flora of Ecuador during the years 1987–89 see G. Harling, this volume.

Field work

Staff and students at GB have from 1987 to 1989 continued the series of expeditions carried out to Ecuador and other neotropical countries. The field work undertaken has been directly related to the revisions of certain plant groups. The collections of U. Molau and B. Eriksen in Ecuador and Peru 1987–1988 were concentrated on *Bartsia*, *Monnina*, and *Valeriana*, the collections together with Margit Fredrikson also on Orchidaceae. From the general collections, however, several new species have already been described (e.g. *Rumex lepto-*

caulis Brandbyge & Rech. f.). Jan-Eric Bohlin, B. Ståhl and M. Neuendorf, 1987, mostly collected Nyctaginaceae, Theophrastaceae, and *Bomarea*. G. Harling, 1988, primarily collected Compositae.

Research programme

1. TAXONOMY

a. Flora of Ecuador

Cyclanthaceae (Gunnar Harling) - published.
Bixaceae, Cochlospermaceae, Elatinaceae (Ulf Molau) - published.
Scrophulariaceae, *Calceolaria* (Ulf Molau) - published.
Musaceae (Lennart Andersson) - published.
Marantaceae (Lennart Andersson & Mats Hagberg) - published.
Amaranthaceae (Uno Eliasson) - published.
Alstroemeriaceae (Magnus Neuendorf).
Centrospermae (excl. Nyctaginaceae) (Uno Eliasson).
Clethraceae (Claes Gustafsson).
Compositae-Mutisieae (Gunnar Harling).
Nyctaginaceae (Jan-Eric Bohlin).
Papaveraceae (Magnus Lidén).
Polygalaceae (Bente Eriksen).
Rubiaceae (Lennart Andersson).
Theophrastaceae (Bertil Ståhl).

b. Revisions

Calceolarieae (Scrophulariaceae; Ulf Molau) - published, Flora Neotropica.
Bartsia (Scrophulariaceae; Ulf Molau) - in press, Opera Botanica.
Clavija (Theophrastaceae; Bertil Ståhl) - submitted, Opera Botanica.
Alonsoa (Scrophulariaceae; Fanny Astholm).
Bomarea (Alstroemeriaceae; Magnus Neuendorf).
Colignonia (Nyctaginaceae; Jan-Eric Bohlin).
Erioderma (Pannariaceae; Lars Arvidsson).
Leptogium (Collemataceae; Marie Lindström).
Marantaceae (Lennart Andersson).
Monnina (Polygalaceae; Bente Eriksen).
Monotagma (Marantaceae; Mats Hagberg).
Neèa (Nyctaginaceae; Jan-Eric Bohlin).
Rubiaceae (Lennart Andersson, Claes Persson).
Sphaeradenia (Cyclanthaceae; Roger Eriksson).
Sticta (Lobariaceae; Mats Lindqvist).

Staff and students at the Department of Systematic Botany, GB

1. PERMANENT STAFF

G. Harling, professor emeritus, taxonomy of Cyclanthaceae, Musaceae, and Compositae-Mutisieae; phytogeography.

L. Andersson, professor, taxonomy of Scitaminae, especially Marantaceae and Musaceae, and taxonomy and evolution of neotropical Rubiaceae.

U. Eliasson, curator of the Botanical Museum, taxonomy of Centrospermae, especially Amaranthaceae; Myxomycetae.

Magnus Lidén, taxonomy of Papaveraceae, especially *Corydalis;* theoretical systematics.

U. Molau, taxonomy of the Scrophulariaceae; reproductive biology.

M. Neuendorf, curator of the Botanical Garden, Göteborg, taxonomy of *Bomarea* (Alstroemeriaceae).

2. ASSOCIATED STAFF

Lars Arvidsson, serving at the Nature Conservation Unit, City of Göteborg, taxonomy of lichens, especially macrolichens of Ecuador.

3. STUDENTS

a. Ph. D. students

Fanny Astholm, taxonomy and biosystematics of *Alonsoa* (Scrophulariaceae).

Jan-Eric Bohlin, the Nyctaginaceae in the Neotropics: morphology, taxonomy, and intergeneric relationships.

Bente Eriksen, taxonomy of *Valeriana* (Valerianaceae) and *Monnina* (Polygalaceae).

Roger Eriksson, taxonomy of *Sphaeradenia* (Cyclanthaceae).

Sven Fransén, taxonomy of Bartramiaceae (Bryophyta).

Margit Fredrikson, comparative embryological studies in the Orchidaceae.

Mats Hagberg, taxonomy of *Monotagma* (Marantaceae).

Mats Lindqvist, taxonomic revision of the genus *Sticta* (Lichenes).

Marie Lindström, taxonomy of *Leptogium* (Lichens).

Bertil Ståhl, taxonomy of Theophrastaceae.

b. Majoring students

Completed 1989

Claes Gustafsson. A preliminary revision of the family Cletraceae in Ecuador.

Claes Persson. A preliminary evaluation of subfamilial classification of the Rubiaceae.

c. Fellowships in Göteborg:

1987–88 Bente Eriksen, AAU, Polygalaceae. Funded by Nordic Council, research grants.

1989–90 Carmen Ulloa, QCA, Ecuador, montane forests. Funded by Swedish Institute.

Publications on Neotropical Botany 1987–1989

Andersson, L. 1989. An evolutionary scenario for the genus *Heliconia* (Heliconiaceae). – In: L. B. Holm-Nielsen, I. Nielsen & H. Balslev (eds.), Tropical Forests: Botanical Dynamics, Speciation and Diversity. Academic Press, London.

– 1990. The driving force: species concepts and ecology. – Taxon (in press).

– and U. E. Eliasson (eds.) 1987. Botanical studies dedicated to Gunnar Harling. – Opera Bot. 92: 1–282.

Arvidsson, L., Elix, J. A., Mahadevan, I., Wordlaw, J. H. & Jørgensen, P. M. 1987. New depsides from *Erioderma* lichens. – Aust. J. Chem. 40: 1581–1590.

– & Galloway, D. 1990. Studies in *Pseudocyphellaria* II. Ecuadorean species. – Lichenologist 22 (in press).

Bohlin, J.-E. 1988. A monograph of the genus *Colignonia* (Nyctaginaceae). – Nord. J. Bot. 8: 231–252.

Egeröd, K. & B. Ståhl 1990. Revision of *Lycoseris* (Compositae - Mutisiae). – Bot. Gothob. 8 (in press).

Eliasson, U. H. 1987. Amaranthaceae. – In: G. Harling & L. Andersson (eds.), Flora of Ecuador 28: 1–138.

– 1988. Floral morphology and taxonomic relations among the genera of Amaranthaceae in the New World and the Hawaiian Islands. – Bot. J. Linnean Soc. (London) 96. 235–283.

Eriksson, R. 1989. *Chorigyne*, a new genus of the Cyclanthaceae from Central America. – Nord. J. Bot. 9: 31–45.

– 1990. The genera *Sphaeradenia* and *Stelestylis*. – In: Flora of the Venezuelan Guyana (in press).

Fransén, S. 1988. On the status of *Bartramia campylopus* Schimp. in C. Müll. and *Gymnostomum setifolium* Hook. et Arnott. – Lindbergia 14: 30–32.

Kennedy, H., L. Andersson & M. Hagberg. 1988. Marantaceae. – In: G. Harling & L. Andersson (eds.), Flora of Ecuador 32: 1–191.
Molau, U. 1988. Scrophulariaceae – Part I. Calceolarieae. – Flora Neotropica Monograph 47: 1–326. New York.
– 1990. The genus *Bartsia* (Scrophulariaceae-Rhinanthoideae). – Opera Bot. (in press).
Sparre, B. & Andersson, L. A taxonomic revision of the Tropaeolaceae. – Opera Bot. (in press).
Ståhl, B. 1987. The genus *Theophrasta* (Theophrastaceae). Foliar structures, floral biology and taxonomy. – Nord. J. Bot. 7: 529–538.
– 1989. A synopsis of Central American Theophrastaceae. – Nord. J. Bot. 9: 15–30.
– 1990. Theophrastaceae. – In: G. Harling and L. Andersson (eds.), Flora of Ecuador 39: 1–21 (in press).
– 1990. Primulaceae. – In: G. Harling and L. Andersson (eds.), Flora of Ecuador 39: 23–35 (in press).
– 1990. Theophrastaceae. – In: Flora Mesoamericana (subm.).
– 1990. Theophrastaceae. – In: Flora de la real expedición botánica de nuevo reino de Granada (in press).
– 1990. A revision of *Clavija* (Theophrastaceae). – Opera bot. (subm.).

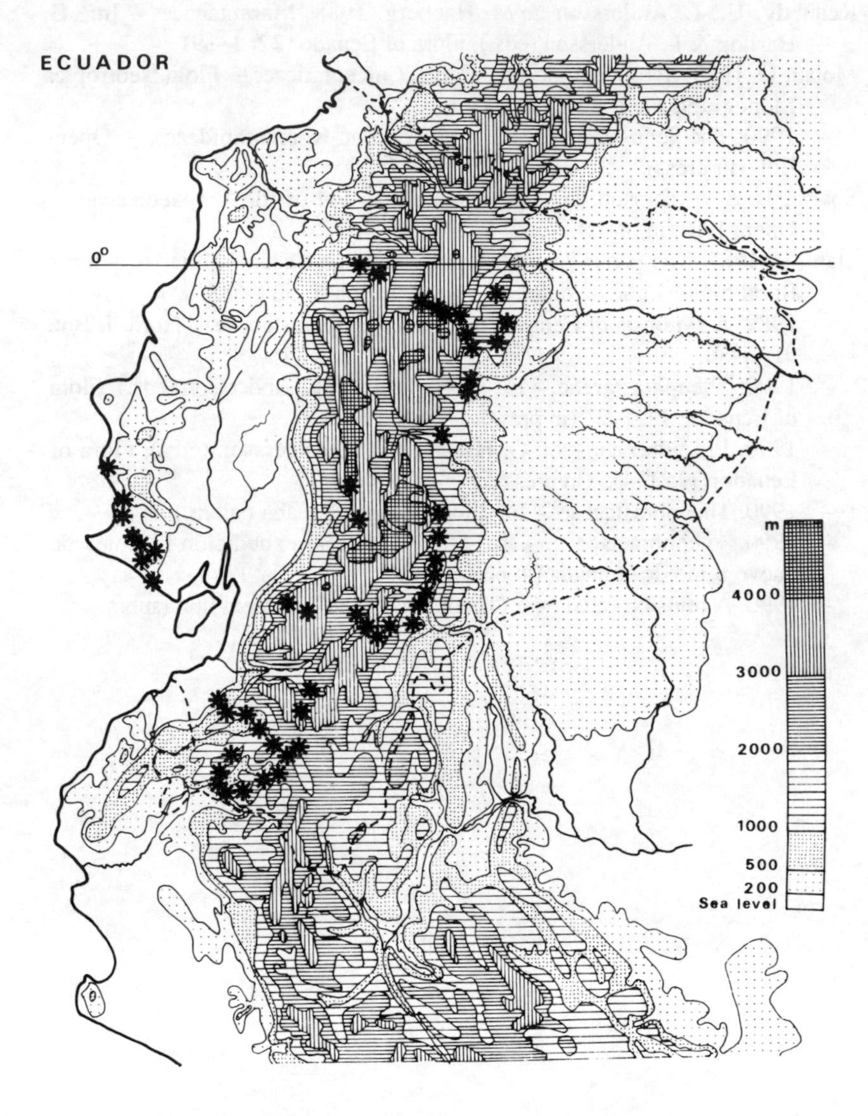

Collection localities of Jan-Eric Bohlin, Bertil Ståhl, Roger Lundin and Magnus Neuendorf nos. 1100--1573, 1987

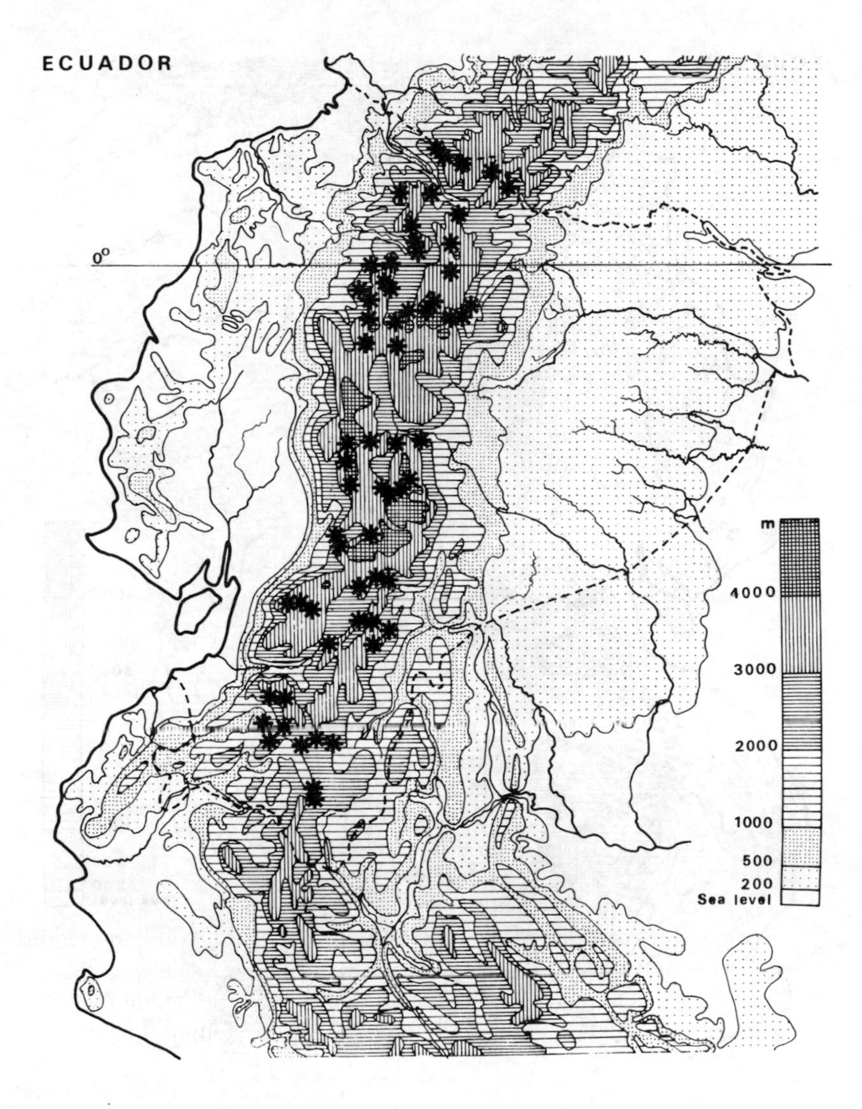

Collection localities of Ulf Molau & Bente Eriksen nos. 2001--3300, 1987--1988, with the assistance of Margit Fredrikson and Birgit Malm.

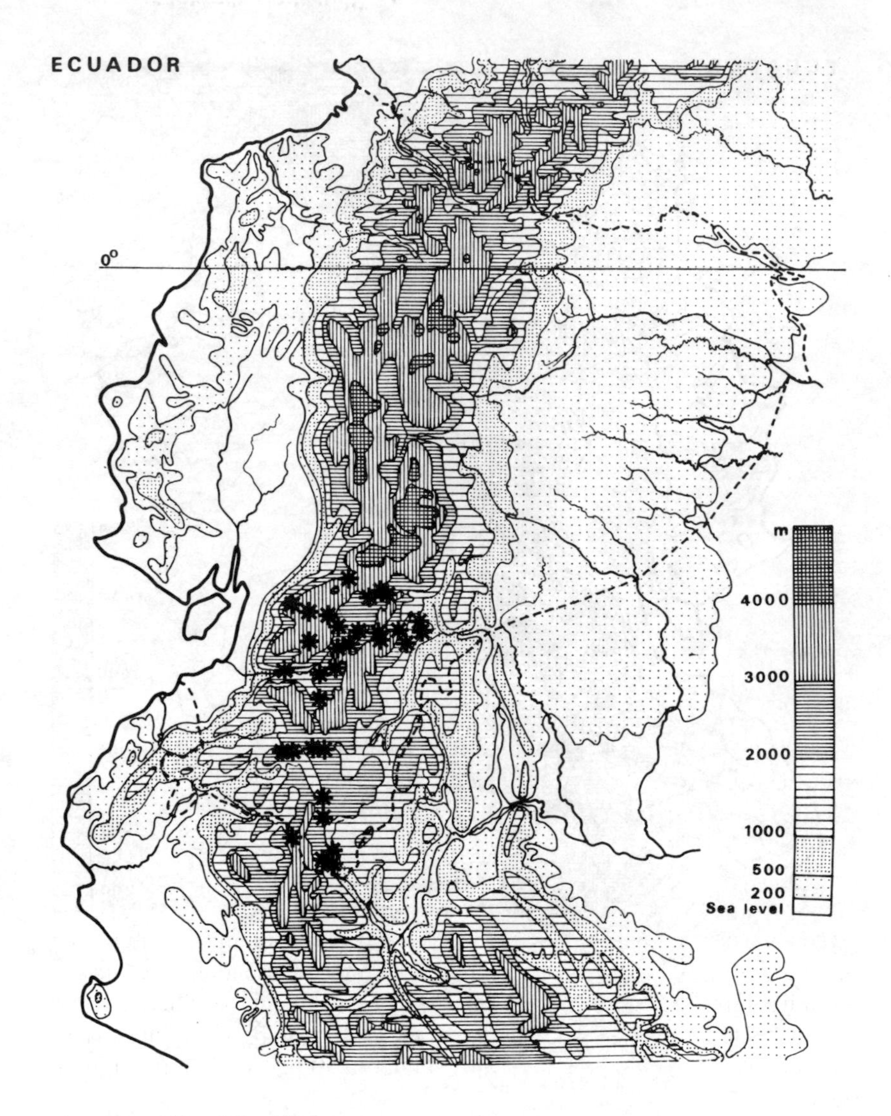

Collection localities of Gunnar Harling nos. 25101--26206, 1988

Concluding remarks: Nordic botanical research in the Andes and western Amazonia

Lauritz B. Holm-Nielsen

Rector of The Danish Research Academy, Paludan Müllers vej 17, DK 8000 Aarhus C, Denmark

Thirtyfour scheduled presentations in two days. The programme for this symposium, the first Nordic symposium on the botany of South America, which follows the Aarhus and Göteborg workshops in 1984 and 1986, has really been a tight one, comnprising such diverse disciplines as taxonomy, floristics, geomorphology, biogeography, plant sociology, biology, ecology, ethnobotany, and history of botanical exploration

Examples from a variety of plant groups and floristic areas have been presented. The studies altogether give a comprehensive survey of what is going on in this important research area, and a good idea of what possibly will happen in the next couple of years.

It has been a pleasure to learn about the progress in projects which were only "mlrages" three years ago. The involvement of Turku and in Bergen is of course the best sign of the vitality of Nordic neotropical botany. For those who have been involved in neotropical botany for decades, it is rewarding to witness all these many activities, and to see how the Nordic scientific community, by joining efforts, can provide a powerful and remarkable contribution to science of tropical areas.

The present research teams will doubtlessly serve as strongholds in future tropical ecology and botany. But the most important contribution from the involved institutions may well be the training of young botanists. There are not many places in the world, where a symposium like this can present such a list of speakers with an average age closer to 30 than to 50 years. This is a sign of strength.

This symposium has, like the preceeding workshops, been an "inhouse" activity in the Nordic house, but it is a pleasure to have Alba, Carmen, Katya, and Tania here, representing the neotropical countries.

List of all participants in the symposium:
Those marked with * have contributed to this volume

University of Bergen
Cornelius C. Berg*

University of Göteborg
Lennart Andersson*
Jan-Erik Bohlin
Roger Eriksson*
Margit Frederiksson*
Mats Hagberg
Gunnar Harling*
Jette Knudsen*
Magnus Lidén
Claes Persson
Bertil Ståhl*
Lars Tollsten*
Carmen Ulloa*

University of Stockholm
Roger Lundin*

University of Turku
Risto Heikkinen
Risto Kalliola*
Yrjö Mäkinen*
Maarit Puhakka*
Tania de la Rosa*
Kalle Ruokolainen
Jukka Salo*
Hanna Tuomisto

University of Copenhagen
Ole Seberg*

Aarhus University
Alba Luz Arbeláez*
Henrik Balslev*
Ulla Blicher-Mathiesen*
Finn Borchsenius*
John Brandbyge*
Dorte Christensen
Henning Christensen
Lars Christensen*
Steven Churchill*
Lis Elleman
Bjarke Eriksen
Lauritz B. Holm-Nielsen*
Peter Møller Jørgensen*
Bente Bang Klitgaard*
Lars Peter Kvist*
Simon Lægaard*
Eva Bjerrum Madsen*
Ingvar Nielsen*
Anders Pedersen
Axel Dalberg Poulsen*
Katya Romoleroux*
Susanne S. Renner*
Benjamin Øllgaard*